B-58 HUSTLER

The Greased Lightning of Strategic Air Command

H.J. CAMPBELL

Electrikbooks

On the cover: Convair B-58A (59-2456) in flight with multiple weapons loaded during testing April 1961.

Table of Contents

Introduction

In October 1963 a B-58A, SAC's "Greased Lightning", set five world records including a speed record flying 8,028 miles from Tokyo to London in 8 hours, 35 minutes, and 20.4 seconds at 938 mph average. This bested the previous record of 17 hours and 42 minutes held by a British jet by more than half. The flight demonstrated SAC's ability to divert a B-58 to hit a target at any spot on the globe and do it supersonically!

But this wasn't the first record for the B-58. It set more records than any other combat aircraft and still holds the Tokyo to London speed record. It set 19 world records including 13 speed records, one endurance, three payload, and two altitude records with the aircraft flown in combat configuration by SAC combat crews. Five of the records were formerly held by the Soviet Union. The B-58 won all major international awards including the Bendix Trophy, Thompson Trophy, Blériot Trophy Cup, Mackay Trophy, and Harmon Trophy.

Designed in the 1950s, the B-58 Hustler provided the U.S. Air Force with a capability not available in any aircraft in its day. It is the only Mach 2+ supersonic bomber ever produced. It could fly at high altitudes or at low-level just below the speed of sound, carrying five nuclear weapons to its targets. It was nearly undetectable by enemy radars due to its low radar cross section and was better than larger aircraft at flying through heavy turbulence due to its solid delta wing design. It could get airborne in only half the time required by other bombers. Because of its ability to take off very quickly there was no need to spread the Hustlers over many bases keeping operating costs low.

The B-58 quickly became a very effective aircraft. By the end of 1963, Air Force crews had made over 10,500 flights in the B-58, amounting to about 53,000 flying hours. Of these hours, 1,150 were supersonic and 375 were at Mach 2.0. By the mid-1960s, with the B-47 nearly phased out, the B-58 and B-52 were the only two bombers programmed into the 1970s. They formed the all-important mix of aircraft and missiles which Air Force leaders unanimously agreed was critical in maintaining a credible deterrent force.

Despite its growing acceptance, on 8 December 1965 Secretary of Defense Robert McNamara directed the phase out of 345 older B-52C through F models and all B-58 bombers by June 1970. On 10 December, McNamara also announced the decision to procure a new FB-111A aircraft to replace the retiring B-52 and B-58 aircraft. These decisions were met with objections and concerns from Air Force leaders and many in Congress. Nonetheless, after several changes and delays in the retirement plans, the last of the B-58 fleet arrived at the boneyard six months early on 16 January 1970.

But the B-58 was loved by many and left behind a legacy of a highly advanced aircraft that employed the latest technologies and design techniques, and the only Mach 2+ bomber ever to serve in U.S. Air Force operational squadrons!

Design and Development

The B-58 Hustler was a truly remarkable aircraft that provided the U.S. Air Force with a capability not available in any aircraft in its day. Designed in the 1950s, it is the only Mach 2+ supersonic bomber ever produced. It could fly at high altitudes or at low-level just below the speed of sound, carrying five nuclear weapons to its targets. It was nearly undetectable by enemy radars due to its low radar cross section and was better than larger aircraft like the B-52 at flying through heavy turbulence due to its solid delta wing design. It had a highly accurate Bombing-Navigation System (BNS) and could get airborne in only half the time required by other bombers. Because of its ability to take off very quickly there was no need to spread the Hustlers over many bases keeping operating costs low. By the end of 1963, Air Force crews had made over 10,500 flights and logged 5,300 hours (1,150 of them supersonic, including 375 at Mach 2).[1] Despite initial teething troubles the B-58 had become an effective weapon system by the mid-1960s.

The B-58 Hustler's delta wing design was critical to its ability to fly at Mach 2+ speeds. The first successful delta wing designs were explored in Germany before World War II

B-58A (59-2433) Hustler in flight. The Hustler's delta wing design was critical to its ability to fly at Mach 2+ speeds (National Museum of the U.S. Air Force)

(WWII) by Alexander M. Lippisch, an eminent German scientist, who is credited with delta wing development. However, although Lippisch created the delta wing concept he was not the only contributor to its development. He followed a path first taken by John Dunne in the United Kingdom (UK) who developed such aircraft prior to World War I (WWI). Lippisch's efforts also paralleled those of G.T.R Hill in the UK and John Northrop in the United States (U.S.) whose flying wing concepts led to the development of the delta wing configuration. Many U.S. agencies, including the National Advisory Committee for Aeronautics (NACA), studied Lippisch's captured reports after WWII although NACA had already explained many of the delta wing theoretical advantages by 1945 independent of Lippisch.

Major General Curtis LeMay, who would later become the renowned "father" of Strategic Air Command (SAC), was assigned as the first Army Air Forces (AAF) Deputy Chief of Staff (DCS) for Research and Development (R&D) in 1946 after a brief assignment at Headquarters Air Materiel Command (AMC) following WWII. He worked very closely in his new role with Theodore von Karman, AAF Chief Scientific Advisor, a Hungarian-born aerodynamicist. Both LeMay and General Henry "Hap" Arnold, Commanding General of the AAF, had great respect for von Karman. LeMay quickly established a diverse R&D program, which included the Bell X-1 rocket plane, and was probably influenced in part by von Karman's prediction in 1945 that future aircraft "will move with speeds far beyond the velocity of sound."

As one of LeMay's first R&D programs, the AAF awarded a contract to the Consolidate-Vultee Aircraft (Convair) Corporation[2] in October 1946 for a Generalized Bomber (GEBO) study. The AAF called for the determination of design trends for supersonic flights including low aspect wings and delta wings and effectively marked the beginning of B-58 development. Convair had already previously studied Lippisch's reports and various wing configurations including swept and semi-delta wings before concentrating on delta wing designs for supersonic flight. Convair studied several configurations to determine the effects of wing area, aspect ratio, thickness and sweep, as well as the impacts on speed, range, and gross weight of the engine type (turbojet or turboprop), size, and number installed.[3]

LeMay wrote a letter in 1947 to Lieutenant General Nathan F. Twining, AMC Commander and later Air Force Chief of Staff, and urged a new jet bomber development to be operational in the late-1950s. By October 1947 LeMay had moved on to become Commander, United States Air Forces in Europe (USAFE), but the Air Force issued new medium bomber requirements reflecting his letter to Twining. Air Force engineers at AMC wanted realistic requirements although the Air Staff believed that mission objectives should drive technological development.[4] The Air Staff requirements included a 2,500-mile combat radius, 500 mph cruising speed, and 170,000 pounds gross weight.

Alexander M. Lippisch

Alexander Lippisch leans over the nose of the Storch V in 1929. (German Federal Archives)

Alexander Lippisch was a German aeronautical engineer who developed and conceptionalized delta wing designs which functioned practically in supersonic delta wing aircraft. Lippisch was born in Munich and developed an interest in aviation after seeing a demonstration by Oriville Wright at Templehof Aerodrome in September 1909. Between 1915 and 1918 he served in the German Army as an aerial photographer and mapper during WWI. This increased his interest in aviation and he decided to abondon his previous interest in art school.

Following the war he worked for the Zeppelin Company and developed an interest in tailless aircraft. His first design, in 1921, was built by Gottlob Espenlaub and designated the E-2 glider. This was the beginning of a research program that resulted in about 50 designs during the 1920s and 1930s. He was appointed in 1925 to Director of the Rhon-Rossitten Gesellschaft (RRG), a glider organization including research groups and construction facilities. Beginning in 1927, he resumed his tailless research with a series of mostly glider designs called Storch I – IX. He also began exploring the possibility of rocket propulsion by designing a series of rocket-propelled gliders with wingspans of seven feet or more weighing in excess of 30 pounds. On 11 June 1928, his rudimentary design called the *Ente* (Duck) became the world's first successful manned, rocket propelled arcraft.

Lippisch followed the Storch series with a series of tailless delta designs which were the earlest delta wing designs ever developed. In 1931, Delta I was the first to fly followed by Delta II and III. The Delta IV was built in four powered variants designated IVa through IVd. Delta V was a blended body delta wing design. These designs were built between 1931-39.

In 1939, Lippisch moved to the Messerschmitt plant in Augsburg to work on high speed fighter aircraft designed around rocket engines. Wind tunnel tests suggested that the delta wing was a good choice for supersonic flight and Lippisch set to work on designing a supersonic, ramjet powered fighter. He continued to work on this throughout WWII. However, by the time the war ended the project had only advanced as far as a developmental glider.

Lippisch was taken to the United States after the war, like many German scientists, under Operation PAPERCLIP. Consequently, his research papers and designs were captured by the U.S. Government and made available for study by American companies. He also worked for the U.S. Air Force at White Sands Missile Range where he continued to advance delta wing design theories.

Following his time with the Air Force, Lippisch worked for the Collins Radio Company, Aeronautical Division, from 1950-64. He shifted his attention to ground affect craft, designed to glide over flat surfaces using the interaction between the moving wing and the surface below. However, he soon developed cancer and resigned. When he recovered in 1966, he formed his own company called the Lippisch Research Corporation and built prototypes of his designs.

Lippisch also conceived of a Vertical Take-Off and Landing (VTOL) craft which he called an "aerodyne" during his time at Collins. Its fuselage comprised a large ducted rotor, and the thrust could be varied between downward for vertical takeoff and landing, and backward for forward flight. He worked principally with both Collins and Dornier Aerodyne in its development. But neither type got beyond the prototype stage. Lippisch died in Cedar Rapids, IA on 11 February 1976. He was inducted into the International Air & Space Hall of Fame at the San Diego Air & Space Museum in 1985.

The B-50 and B-36 could not be modified to meet these requirements and LeMay believed a new strategic bomber should follow-on the B-52 then in development.

Meanwhile, aviation technology was progressing rapidly. The Bell X-1 rocket plane completed its first supersonic test flight on 14 October 1947 ending speculation that "aerodynamic forces became infinite at Mach 1."[5] Convair continued delta wing development. Although not the inspiration, Lippisch's report reinforced and encouraged Convair engineers to believe that the delta wing could solve most of the problems of supersonic flight. The Convair XF-92, forerunner of the F-102 Delta Dagger, was the winning interceptor design submitted in 1946. No production contract was awarded for the XF-92, but it was important to delta wing development.[6] Boeing, on the other hand, continued to explore several high-performance turbojet developments to meet the new medium bomber requirements. The first Boeing XB-47 development contract was issued in December 1945 and in October 1948 Boeing presented its first XB-52 design using Pratt & Whitney (P&W) J-57 turbojets in place of the turboprops previously proposed.

The Convair XF-92, forerunner of the F-102 Delta Dagger, was important to delta wing development. (U.S. Air Force)

The Convair GEBO study was complex, but still the Air Force contracted to Boeing for a second bomber study in October 1947. Requirements were firming up and the Air Force requested a medium bomber weighing less than 200,000 pounds with a 2,000-mile radius capable of carrying a 10,000-pound bombload and complete all around defensive armament. Convair recommended a 100,000-pound delta wing design operating at Mach speeds. Boeing recommended a conventional turboprop design which eventually led to the Boeing XB-55 medium bomber. The different recommendations between Convair's high-speed bomber and Boeing's conventional design caused opposing concepts to emerge among Air Force leaders.

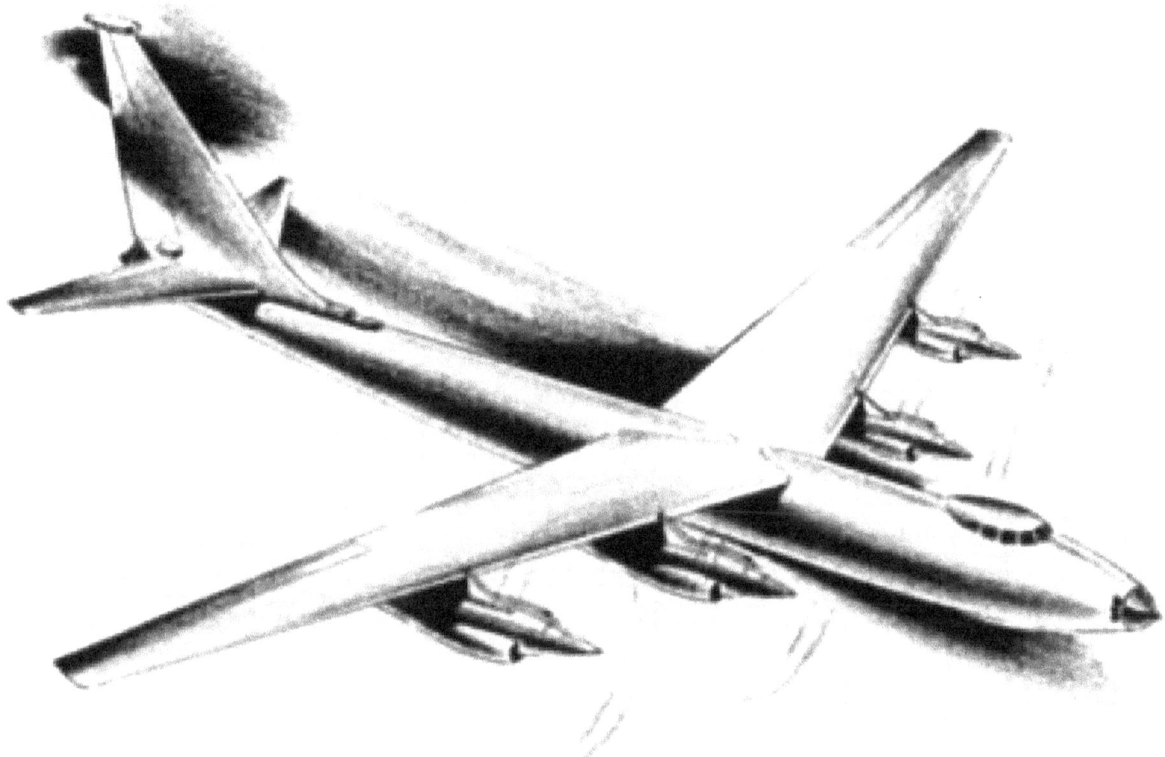

Boeing recommended a conventional turboprop design which eventually led to the Boeing XB-55 medium bomber. (Unknown)

By June 1948, Convair completed its GEBO study which was documented in three reports. Convair had studied 10,000 different configurations which explored the effects of different wing areas, aspect ratios, thickness, and sweep. In addition, several different types of propulsion systems were considered, both turbojet and turboprop. The final report entitled *Generalized Bomber Research Turboprop Airplanes* prompted the Air Force Aircraft Projects office to host a three-day symposium of Government and major contractor representatives beginning on 18 August 1948. The symposium was called to survey the latest developments in turbine and propeller propulsion and to determine trends in size, weight, and performance related to forthcoming bomber aircraft.

Following the symposium, Convair issued additional reports on powerplant and propeller trends for turboprop aircraft with a 50,000-pound bombload with gross weights of 1,000,000 to 1,200,000 pounds and ranges from 0 to 20,000 miles. Reports covering turbojet aircraft with both 0- and 35-degree wing sweep angles were issued in December. Consequently, Air Force engineers at AMC asked for a continuation of the GEBO studies in late 1948 to show the "feasibility of military characteristics" and establish "balanced characteristics and desirable design compromises." This included studying the effects of small changes in such characteristics as drag, specific fuel consumption, propeller efficiency, and engine weight on the aircraft shown in the study, thereby improving the usefulness of the basic aircraft data covered in the GEBO contract.

Testing successes of Convair's XF-92 delta wing, which first flew in early 1949, quickly attracted attention within the Air Force. Secretary of the Air Force (SECAF) Stewart Symington, and other Air Force leaders, were searching for "new and imaginative solutions." Brigadier General Donald Putt, Director of the R&D Office, DCS Materiel, outlined AMC's many future tasks for industry and government. Among them were the continuation of B-52 development and an effort to develop a possible supersonic bomber. He further recommended that AMC solicit the industry for "new and possible unconventional approaches to the intercontinental bomber problem on the basis of minimum specific military requirements to be established in cooperation with this Headquarters."

The Boeing XB-55 medium bomber was cancelled on 27 January 1949 due to a lack of funds and longer than expected development time. The Air Force also determined that there was substantial XB-47 progress, and there was no longer an immediate need for another subsonic medium bomber. They believed the design should have been "predicated on greater aerodynamic achievements and improved propulsion."[7] However, the XB-55 did have some significant developments, and the Air Force cancelled the contract in a way that allowed maximum benefit such as retaining study reports and increasing the tunnel tests scope.

Meanwhile, as Air Force leaders were struggling to define and fund the nebulous supersonic bomber requirements, a conceptual supersonic strategic bomber was emerging from the lower elements at AMC. The Heavy Bomber Committee of the Aircraft and Weapons Board recommended that the total distance from base to target for the strategic bomber be separated into three zones which would consist of logistics, combat, and target. The outer zone was considered the combat zone which included the target zone. The inner zone (nearest to base) was considered the non-combat zone, and it was considered a "logistical problem" to get the aircraft to the combat zone. Aerial refueling would be used inside the non-combat zone to extend range.

The objective of this zoning would be to improve target area performance and led to the idea of supersonic target zone performance. Among the unconventional concepts which evolved to accomplish supersonic combat and target zone capability were the use of supersonic parasites in combination with larger logistics zone carriers; an inflight refueled medium bomber with an adequate escape radius; the use of drones; and the employment of a medium bomber with exceptionally high target zone performance.[8]

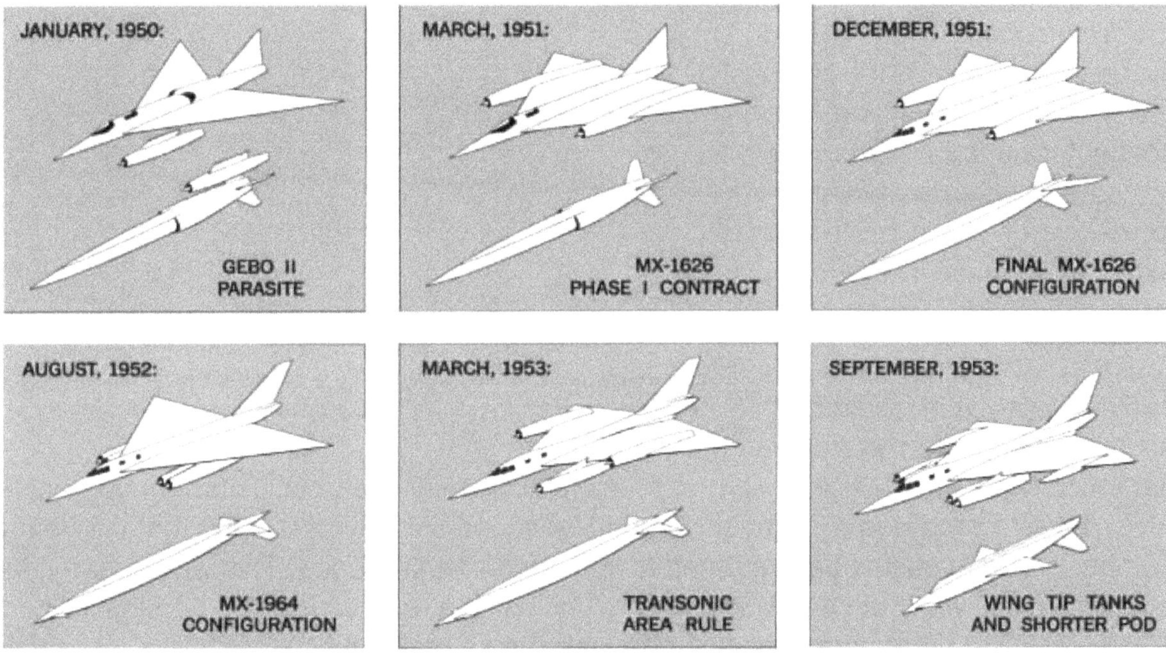

B-58 design evolved through several configurations starting with the GEBO II study which was based on a parasite concept that would be released and retrieved by a B-36 bomber. (Unknown)

Consequently, the Air Force issued contract AF33(038)-2664 for a new Convair GEBO II long-range supersonic bomber development study on 6 June 1949. Initially, Convair was asked to settle the turboprop vs. turbojet powerplant controversy, and to apply the solution to a bomber with a radius of 1,200 to 2,500 nautical miles with a 10,000-pound bombload, cruising speed of over 450 knots, flight altitude above 35,000 feet,[9] and take-off distance less than 6,000 feet. Convair was also required to explore in detail the performance characteristics of high-speed medium bombers, explore the effects of different equipment characteristics, and incorporate other useful suggestions from the first GEBO symposium.

Convair initially proposed a 100,000-pound, small delta wing aircraft coded FZP-110 in January 1950 with two crewmen, a two-stage system (release/retrieval), and four General Electric (GE) J53 turbojet engines without afterburners (one in the tail, two droppable under the wings, and one in the tail of a long, finned droppable bomb pod). The aircraft would be transported to the combat zone as a parasite under a B-36. The parasite would cruise to the target at Mach 1.3 and then sprint over the target at Mach 1.6 and reach a maximum altitude of 48,500 feet before pod release. Radius was estimated at 3,000 miles. After the attack the "return component" would fly back to the B-36 at Mach 0.9 with its single engine. The concept included no defensive armament since the high performance of the aircraft was expected to make it immune to interception.

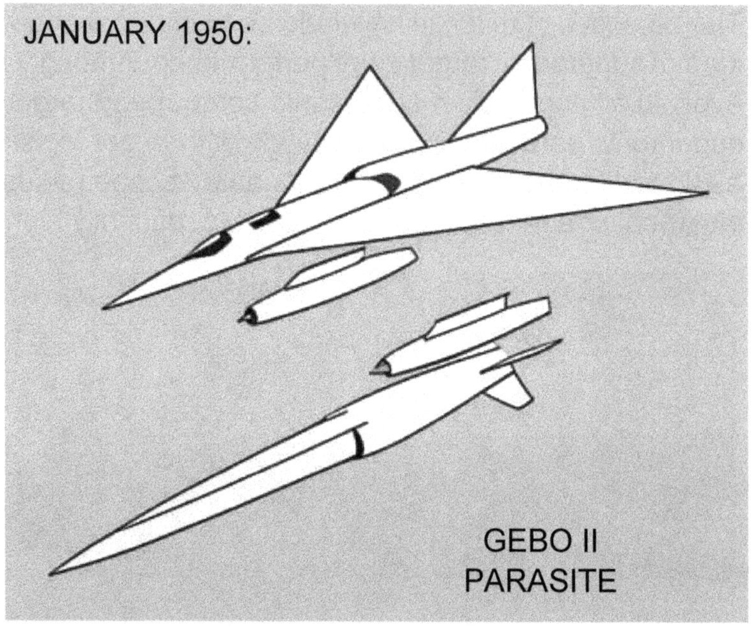

Convair initially proposed in January 1950 a small delta wing parasite aircraft that would be carried to the combat zone under a B-36 bomber. (Unknown)

A RAND study in the early 1950's stated that minimizing aircraft size would reduce radar reflectivity. This was clearly a plus for Convair's small supersonic bomber approach. Also, the Air Force Heavy Bomber Committee issued a decision in 1950 that the heavy bomber program should be expanded to include reconnaissance. This was already becoming typical, with both the B-36 and B-47 including a reconnaissance version. In April 1950, the GEBO II requirement was changed based on Convair's parasite proposal to require a radius of 3,000 to 4,500 nautical miles from the operating base with speeds of Mach 0.9 to 1.5 in the 500- to 2,000-mile-deep target zone. This new study resulted in several configurations.

Convair subsequently modified its proposal in the fall of 1950 to a 100,000-pound aircraft with five engines. Two fixed jet engines were buried in the wings, two droppable engines were under the wings, and a fifth engine was in the long, streamlined pod. The return component included a tricycle landing gear for landing at home, which eliminated the need for retrieving the parasite after the mission. Once airborne, the B-36 would carry the parasite 2,000 miles toward the target, release it and return to base. The parasite would then use its five engines to accelerate to a Mach 1.3 cruising speed. The speed would increase to Mach 1.5 when the parasite was over the target. The parasite would then release the pod containing the nuclear warhead and return home at Mach 1.3. Once safely out of the combat zone, the aircraft would drop its two wing-mounted expendable engines and complete the trip back home at Mach 0.9.[10]

The aircraft would have no defensive armament, but it would be equipped with a full suite of electronic equipment including an advanced BNS and passive Electronic

Countermeasures (ECM) system. Originally the pod was intended to carry fuel and conventional explosives, but early in the GEBO II program a single thermonuclear warhead was incorporated. The long-finned and streamlined bomb pod with a jet engine in its tail could function as a free-fall bomb, with small wings as a powered glide bomb, or an air launched missile, and as late as 1953 it was expected that the pod, with propellant system and fins, would have a range of 100 nautical miles.[11]

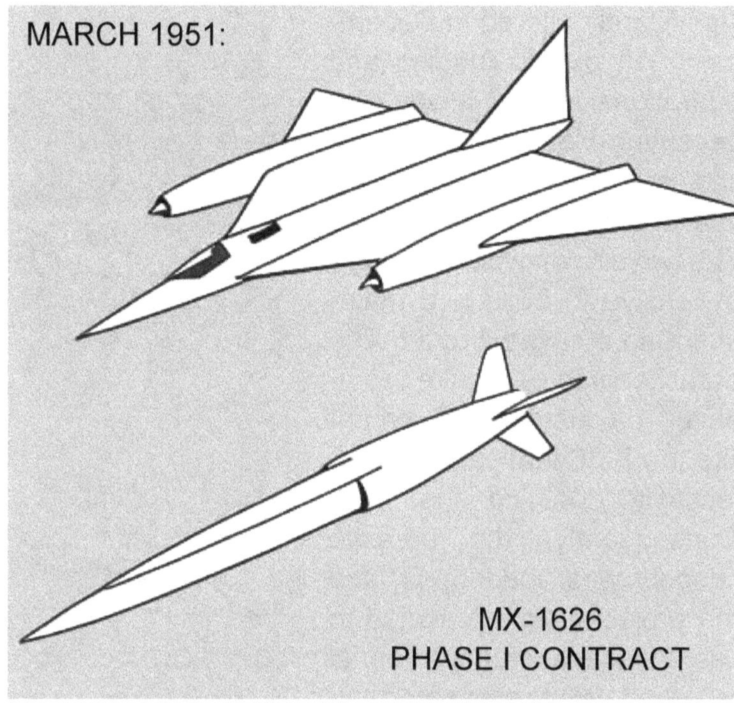

Convair presented a new Phase I proposal that was accepted by the Air Force as Project MX-1626 with the final concept set in March 1951. (Unknown)

On 26 January 1951, Convair presented a new proposal for Phase I development (initially without a mock-up) of a long-range supersonic reconnaissance bomber. This proposal was accepted by the Air Force as Project MX-1626 and letter contract AF33(038)-21250 was signed on 17 February 1951. The final concept was set in March 1951, and the initial flight test date was set for late 1954. Convair's previous GEBO II designs were modified to use only three GE J53-GE-X25 engines with two installed in wing nacelles and one in the droppable pod. The design continued to rely on the parasite approach using a 107,000-pound, delta wing, reconnaissance bomber. Maximum speed over the target was anticipated at Mach 1.7 and the radius was estimated as 4,600 miles. The delta wing had an area of 1,200 square feet and a total span of 47 feet. The fuselage was 82 feet long and included a triangular-shaped vertical tail. As with the previous configuration, the return component was equipped with landing gear for landing at home after the mission.

Boeing also presented a proposal for Phase I development of two reconnaissance bombers through wind tunnel testing, engineering, and mock-up. They proposed the Boeing Model 484-405B, which was a 200,000-pound medium range bomber with conventional 47-degree sweptback wings capable of supersonic flight for part of the mission. The aircraft could drop both atomic and conventional bombs from altitudes of 45,000 to 50,000 feet and had a radius of 2,500 miles. It would fly at Mach 0.9 and increase to Mach 1.3 within 200 miles of the target. It was powered by four P&W J57-P-5 engines with afterburners. The thin wings used to achieve supersonic flight required all fuel in the fuselage. The aircraft was capable of aerial refueling, and extended wing tips could be installed, to extend the range for longer missions. The bomb bay was similar in size to the B-47. After reviewing several different Boeing configurations this was accepted by the Air Force as Project MX-1712 and letter contract AF33(038)-21388 was signed on 26 February 1951. The initial flight test date was also set for late 1954.

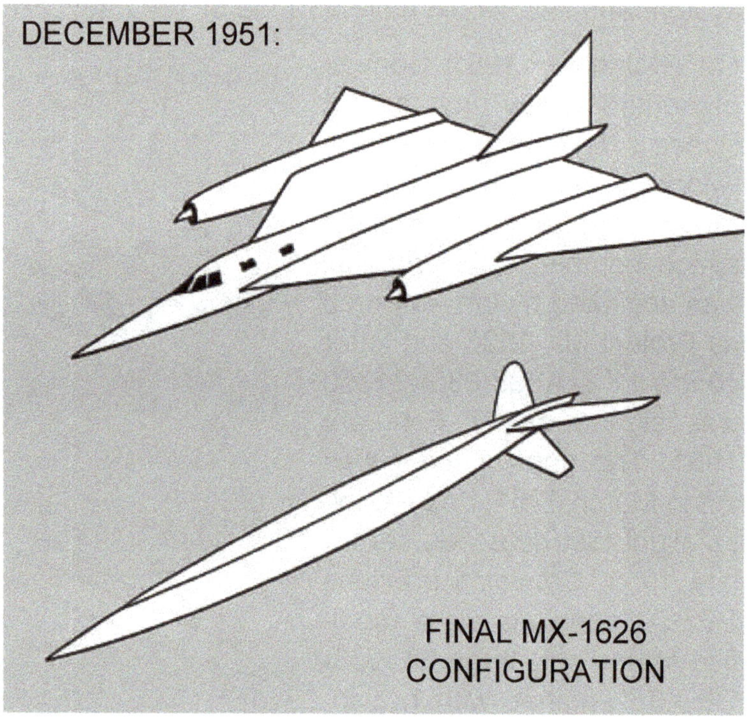

The parasite concept was abandoned in December 1951 due to funding and operational issues. (Unknown)

Convair Project MX-1626 was significantly altered in December 1951 due to funding and operational issues. It was now becoming clear that the parasite concept was financially and operationally unworkable. The two aircraft would be more expensive to build and maintain than a single aircraft. The parasite would require complete navigation equipment like the B-36 that carried it, effectively doubling the cost. Consequently, the parasite concept was abandoned, and the project was redesigned to use a single aircraft with aerial refueling which eliminated the need for the B-36 to carry it to the combat zone. The third engine housed in the pod was eliminated in favor of two engines, now with afterburners, embedded in the wings. The landing gear was modified to support 126,000 pounds gross weight take-offs and the crew increased to three including the pilot, navigator/bombardier, and Defensive Systems Operator (DSO).

In January 1952, the Air Force asked Convair to prepare package program costs and tentative delivery schedules for specific numbers of airplanes (25, 50, 100) including development and production for Project MX-1626.[12] The engines were to be provided as Government Furnished Property (GFP). However, in late February the program was nearly cancelled due to lack of funding and only an emergency infusion of $100,000 kept it going temporarily. A supplemental agreement on 15 May modified the 17 February 1951 letter contract and fully funded Phase I development at $2.8 million. In June 1952, Convair proposed moving the engines to nacelles placed underneath the wing. The engines were to be either two J75-P-1s, two J67-W-1s, four J57-P-7s, four J73s or two J77s.

On 1 February 1952, the Air Force issued the *General Operational Requirements (GOR) for a Supersonic Aircraft Bomber (SAB),* also called *GOR No. 8 (SAB-51)*. These requirements were an update to *GOR No. 7*, which was issued on 8 December 1951, and formally added reconnaissance to the previous bomber only requirement. The requirements called for a multi-mission (strategic reconnaissance/bomber) aircraft capable of day and night all-weather operations. The requirements stated that the aircraft must be capable of carrying a 10,000-pound bombload with a 5,000-mile radius. The radius had to be achieved with a single refueling on the outbound flight only. The aircraft had to fly supersonic speeds at altitudes of 50,000 ft or higher, and high subsonic speeds at low-level altitudes. Furthermore, the aircraft had to be easy to fly, highly reliable, and require small air and maintenance crews. The aircraft had to be small to comply with the RAND study that stated minimizing size would reduce radar reflectivity. The aircraft should also use the best ECM systems and active defenses, and the BNS should be as automatic as possible. The aircraft had to achieve production in 5 years, by 1957, and emphasize "economy from the standpoint of cost to our national resources."[13]

The Air Force also quickly issued *Directive No. 34* on 26 February 1952, along with detailed military characteristics, to formalize the GOR into development directives. However, in a significant change General Donald N. Yates, Director of R&D, Office of the DCS for Development, recommended eliminating the requirement for high subsonic speed at low-level altitude. He said it was unreasonable to develop a high-altitude, long-range, supersonic bomber also capable of subsonic low-level flight. Some engineers said it could be done with "proper technological efforts and plenty of money" but many in the Air Staff were not convinced.[14] Therefore, Yates recommendation for a high-altitude only, long range, supersonic reconnaissance/bomber was incorporated into *Directive No. 34*, revised 1 September 1952 as *GOR No. 1 (SAB-52-1)*, after discussion with the Air Council, Air Research and Development Command (ARDC), SAC, RAND, and the Scientific Advisory Board.[15]

On 29 May 1952, the Air Force issued the *Development Planning Objective for Strategic Air Operations during the period 1956-1960*. This document emphasized the need for a small bomber capable of aerial refueling. It was endorsed by the Air Force Council and General Hoyt S. Vandenberg, Air Force Chief of Staff. However, SAC officials, including LeMay, preferred large bombers that could achieve intercontinental ranges without refueling. LeMay reportedly once quipped, "The B-58 is fine as long as you're wanting to bomb Canada." SAC officials held that "high performance alone" could "never ensure mission success" against defended targets and Surface to Air Missiles (SAM).[16] They argued that "even though the best intercontinental bomber available requires some refueling, it does not follow that the optimum system requires a bomber that has no intercontinental capability without refueling." They pointed out that the small bomber's range would prevent it from operating from most Forward Operating Bases (FOB) without refueling. Furthermore, LeMay did not support unconventional designs because he believed they did not foster reliability and economy and might prove operationally unsuitable.

Despite SAC objections to a small bomber, Air Force leaders pressed ahead. Both Convair and Boeing had made considerable progress toward meeting *Directive No. 34* requirements by mid-1952. However, the directive also dictated the use of the new Weapons System Concept, in which the equipment, weapons, electronics and components of the aircraft would be developed as an integrated whole with the contractor having overall weapon system integration responsibility. The new requirements drove a potential for a new competition, and the Air Force considered querying industry again. But AMC had previously requested proposals in 1951 from Douglas, Lockheed, Martin, and North American. Only two of the companies submitted proposals and they were not compelling. Ultimately, the Air Force agreed with Brigadier General John W. Sessums, ARDC Deputy for Development, to continue with development based on the proposals already submitted.

Convair and Boeing were subsequently directed to stop their MX-1626 and MX-1712 studies and begin Phase 1 Design. The Convair MX-1626 was redesignated as MX-1964, and the Boeing MX-1712 was redesignated as MX-1965. Air Force designations XB-58 and XB-59 were tentatively assigned to the two competing projects, even though no production orders were yet forthcoming.[17] According to Major General Donald L. Putt, now Wright Air Development Center (ADC) Commander, contracts for detailed design and mock-ups would be issued in the fall of 1952. The winning design selection was planned for February to March 1953. This increased timeline made the 1957 production start date required by SAB-51 highly unlikely. However, Wright ADC officials speculated that a prototype could fly by that time.

In the summer of 1952, Wright ADC's Weapons Systems Division analysis concluded "no startling discoveries" and that both contractors were meeting performance and size requirements. The Convair MX-1964 design was slightly ahead of the Boeing MX-1965 although extensive development was still required for both designs to determine engine and integrated electronics selections. Despite the previously agreed plans, ADC recommended selecting one of the contractors to move forward to complete the design and mock-up development to save time and money, eliminate simultaneous development issues, and prevent separate electronics/control systems development. Early selection and contract awards could also potentially salvage the 1957 production start date.

The final evaluation of the competing designs in September 1952 left little doubt about the decision. Convair's design was the most promising means of achieving a supersonic capability and met the spirit of the *Development Planning Objective for Strategic Air Operations during the period 1956-1960* and *GOR No. 8 (SAB-51)*, which emphasized minimal size to reduce radar reflectivity. Furthermore, Convair was becoming very familiar with integrated weapon system development techniques based on its XF-92 and F-102 design activities. The design also emphasized supersonic speed and high-altitude, as well as high-subsonic low-level capability. It featured engines mounted in pods dubbed the "Siamese" configuration for the first time. At the time, Wright ADC engineers thought this engine design might be necessary to reduce supersonic drag.

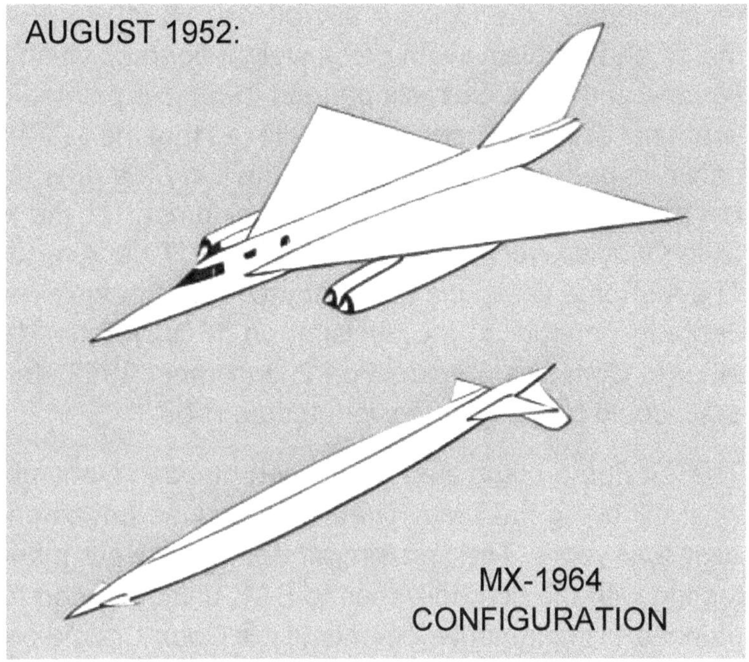

Convair's MX-1964 design was the most promising means of achieving a supersonic capability and left the Air Force little doubt about its decision. (Unknown)

The Boeing design was formally rejected by ARDC on 9 October 1952 because evaluators determined it "would yield either an aircraft of small size with poor supersonic capabilities, or one so large it would not achieve reasonable supersonic performance."[18] The design used conventional swept wings, internal bomb bay, fore and aft bicycle landing gear like the B-52, four GE J73 engines integral within the wing and inboard against the fuselage. It was expected to achieve Mach 1.8 at 55,000 feet, with 156,000 pounds gross weight and ample room for a three-man crew.

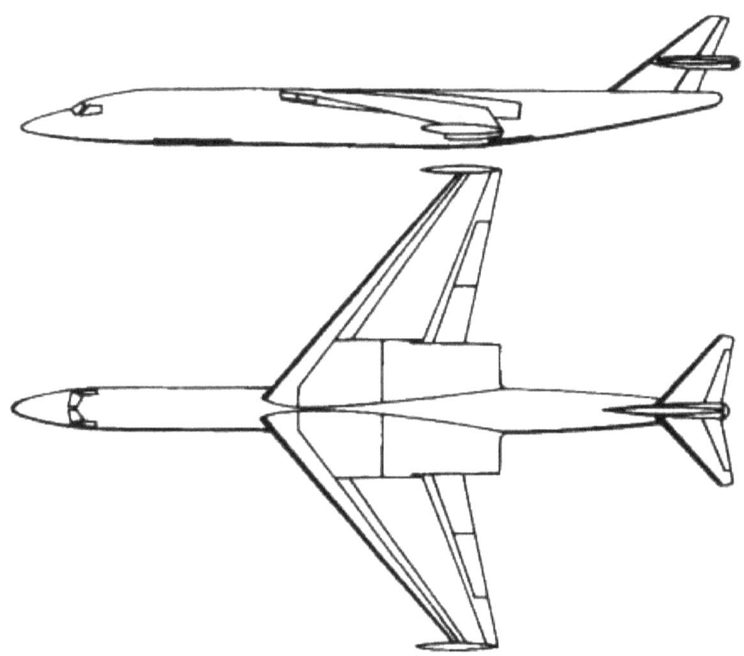

The Boeing MX-1965 was a 156,000-pound medium range bomber capable of supersonic flight. It was powered by four GE J73 engines capable of achieving Mach 1.8 at 55,000 feet. (U.S. Air Force)

Lieutenant General Earle E. Partridge, ARDC Commander, recommended in late October 1952 that the competition between Convair and Boeing be stopped. Partridge recommended the Convair aircraft despite 10 to 15 percent too optimistic drag estimates and problems with gross weight figures, which could potentially limit range, cost figures, and production slippage to 1959. He pointed out that "because of their unconventional design concepts, it is believed that the MX-1964 represents the best balances between performance capabilities, military features, size, and timing for this weapon." He also called for accelerated development of the GE J73-X24A prototype engine, which would eventually evolve into the J79 engine used for production B-58 aircraft. The Air Force Weapons Board approved Partridge's recommendation and General Vandenburg announced the decision on 18 November 1952. Major General Putt wrote a letter to Convair's president on 2 December 1952 stating that as of 5 February the aircraft would officially be designated as B-58.

The Air Force completed production plans in December 1952 even though the design was still being finalized. The schedule was based on the procurement of 244 aircraft over four years. The first aircraft deliveries were planned for January 1956 to support testing with the first deliveries to SAC units planned for January 1958.[19] The Air Force planned to implement elements of the Cook-Craigie production policy[20] and procure the first 30 aircraft for R&D testing while preparations were made for full scale production of a version incorporating all test-dictated changes. All changes made during R&D would be incorporated into the final production configuration and the first 30 aircraft would be retrofitted to the approved configuration. This plan was expected to eliminate faults in the initial design before starting final production.

The Convair Letter Contract AF33(038)-21250, initially issued on 17 February 1951, was modified for the fifth time on 12 February 1953 to begin Phase I development. The B-58 mock-up inspection was scheduled for August 1953, and the RB-58 mock-up inspection was scheduled for fall. The Air Force would manage the program under the new Weapon Systems Development concept which gave Convair overall responsibility for weapons system design and integration. The Air Force planned to monitor Convair plans and progress; approve specifications and subcontractors, supply funding, and veto any developments that could cause operational or logistics problems using a 20-man Joint Program Office (JPO) at Wright ADC made up of ARDC and AMC representatives. The Air Force would supply minimal amounts of GFP and Government Furnished Equipment (GFE) with most supplies and equipment being purchased directly by Convair unless "limitations of industry, operations, or logistics considerations forced the Air Force to control source and/or methodology." All parts would be designed, built, and tested to Convair specifications and Convair would manage several subcontractors to ensure successful systems integration.

The new Weapon Systems Development concept caused some initial misunderstandings and problems. Convair was initially awarded $22 million for production planning and long-lead tools and equipment. However, they expected the contract to also include costs for performing "Weapon Systems Manager" functions as part of its new weapon systems integration responsibilities. The Air Force disagreed and said those costs should be covered in the company's Overhead and General and Administrative (G&A) costs. The Air Force also delayed Convair's $500,000 request to expand the Fort Worth plant to add extra office space for new weapon systems management responsibilities. Ultimately, Convair was forced to spend its own money.

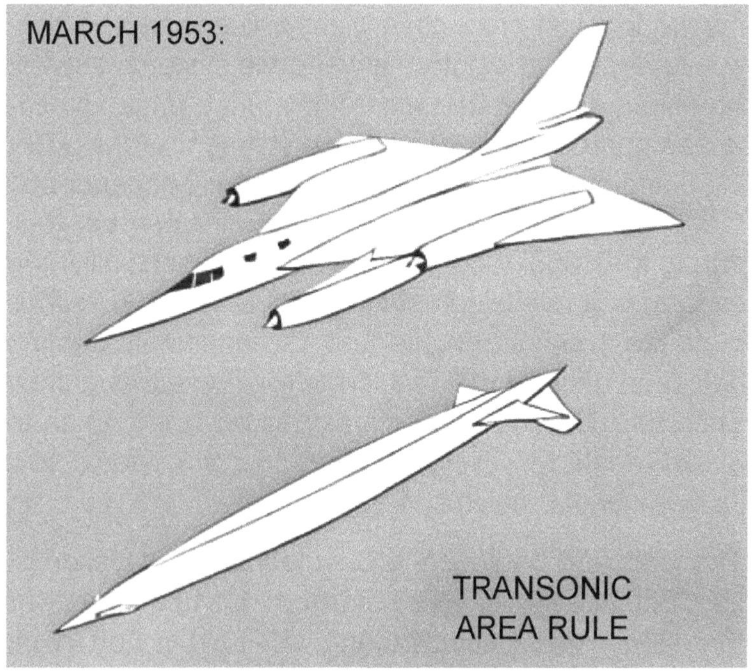

The final B-58 design incorporated the transonic area rule to redistribute the aircraft cross-sectional area longitudinally to minimize drag rise at transonic speeds. (Unknown)

B-58 development continued despite these issues and design approval was given on 24 March 1953. The 60-degree delta wing design incorporated the "area rule concept of aircraft design", also called the Transonic Area Rule, to redistribute the aircraft cross-sectional area longitudinally to minimize drag rise at transonic speeds. This was accomplished by giving the fuselage a "wasp-waist" or "Coke bottle" shape and adding a 10-degree trailing edge angle to increase the wing area to 1,542 square feet. The leading edge was cambered and twisted to reduce drag at lift.[21] The four J79 engines were housed in staggered nacelles with the two inboard units mounted on underwing pylons and the two outboard engines mounted on the wing upper surface. This configuration was chosen to conform to the Transonic Area Rule requirements, even though previous studies indicated the Siamese configuration would entail very little drag penalty.

The Air Force (AMC, ARDC, and Wright ADC) and NACA worked with Convair to refine the design for several more months following the initial design approval. Delays were due to development problems and drag issues with the F-102 delta wing. The F-102 was a growth of the XF-92 delta wing. NACA researcher Richard T. Whitcomb had developed the Transonic Area Rule, which was verified in the Langley Research Center wind tunnel in December 1952. Whitcomb's area ruling was based on the concept that an aircraft would have better supersonic performance if there were as little change as possible in its cross-sectional area along its length, ensuring that airflow did not encounter abrupt transitions in its path along the aircraft. This meant that as the wingspan increased, the fuselage diameter shrank, resulting in the Coke bottle configuration that not only improved performance, but usually improved an aircraft's appearance as well. Although Convair had used the rule to develop the B-58 design approved in March 1953, they resisted finalizing it until it could be verified by their own engineers.[22]

The J79 engine development was also becoming a big issue by May 1953. Air Force leaders were very concerned about J79 development delays. It was destined to become the world's first production Mach 2.0 engine and delays threatened not only the B-58 but several other significant weapons systems such as the F-104, F-4, A3D, and North American A-5 (RA-5). General Partridge emphasized that every effort had to be made to safeguard the successful development of the J79 upon which the "vitally important B-58 and other projects will be so heavily dependent." General Putt told GE executives that the J79 controlled "to a very major degree, this country's ability to defend itself during the 1958-1965 period.... This responsibility should not be treated lightly." However, turbojet development in the U.S. and UK had historically taken 4-5 years to complete the 150-hour engine test. GE engineers confirmed that the engine could not be delivered until July 1957 and the Air Force thought even that schedule was unrealistic. Therefore, the Air Force developed an interim plan to fit the first 18 B-58 aircraft with already-tested P&W J57 engines, but they would soon learn that this solution created its own set of problems.[23]

The first B-58 Engineering Development Inspection of the design, now formally known as Configuration II, was held from 17-18 August 1953. This inspection replaced the formal mock-up inspection and was held in Fort Worth. A second inspection was held on 29 September 1953 to evaluate RB-58 differences from the B-58 configuration. The configuration closely matched the August 1952 MX-1964 configuration and incorporated the requirements contained in *GOR No. 1 (SAB-52-1)*. There were no major systems included (not ready yet) but space was provided. Still, the inspection gave a good idea of the weapon system design. Air Force inspectors, including LeMay, asked for many changes, but none appeared vital. LeMay continued to prefer side-by-side seating for better crew coordination, but this was not possible due to the narrow fuselage, and a jump seat was added as a compromise. Major General Albert Boyd, Wright ADC Commander, summed it up: "It is a radical design, and we must be careful in following through with these technical developments." He added that Convair seemed to have done a very good job.

At this stage in the design, the fuselage of the B-58 still consisted of an upper component and a lower pod that were integral with each other rather than being separated by a pylon. The return component had a flat fuselage undersurface once the disposable pod component had been jettisoned. In addition, nose gear requirements were complicated by the fact that both the pod and the aircraft required nose gear. This meant that several challenges still needed to be addressed as the B-58 design continued.

J79 Engine Development

The GE J79 engine is an axial-flow turbojet engine built for use in a variety of fighter and bomber aircraft including the F-104, B-58, and F-4. (U.S. Air Force)

The GE J79 engine was an axial-flow turbojet engine built for use in a variety of fighter and bomber aircraft and a supersonic cruise missile. In its later versions (J79-GE-17) it had a maximum military thrust of 11,870 pounds and 17,900 pounds with afterburner. The engine was 17.4 feet long and 3.25 feet in diameter. Total weight was 3,835 pounds. It had a single-spool 17-stage axial compressor with variable inlet guide vanes and variable stator vanes, 10 cannular combusters, and a 3-stage turbine. Among its major uses was the Lockheed F-104 Starfighter, Convair B-58 Hustler, McDonnell Douglas F-4 Phantom II, and the North American A-5 Vigilante.

By the late 1940s, jet engine design had progressed to the point where further progress was limited by the performance of its compressor, in particular the pressure ratio of the compressor had to be increased to reduce the engine fuel consumption. However, the useful operating range of the compressor was limited at that time. If designed for high efficiency at high speeds it was very inefficient and prone to stall at low speeds. In 1944, NACA tested a theory for "Extending the useful operating range of axial-flow compressors by use of adjustable stator blades" by running an eight-stage axial compressor and adjustable blade angles. Considerable improvement in efficiencies were obtained at compressor speeds appreciably below the design speed.

By 1950 GE was focusing on supersonic engines with variable stators. Based on their past experience, variable stators promised the best way of designing the compressor for the high required pressure ratio needed to achieve the supersonic performance, subsonic cruise performance, and low weight necessary for future supersonic aircraft. In 1951, a GE team was given funding to build a test compressor with variable stators. In early 1952, the Aircraft Gas Turbine Division lead formed two teams to do design studies for an engine with the flexibility, strength, and thrust to run for extended periods at Mach 2.0 while still giving good fuel economy when cruising at Mach 0.9. One team used a single-spool variable stator configuration, while a parallel effort used the twin-spool approach used successfully by P&W.

Ultimately, the engine with the single-spool variable stator configuration was chosen over the twin-spool approach as it was lighter, simpler, and had a smaller diameter. A new engine, the J73-X24A, had completed testing and proven the technology by May 1952. It produced approximately 14,000 pounds of thrust, with afterburner, to power a new supersonic bomber. The Air Force chose the J73-X24A as the new engine to power the B-58 on 3 December 1952 and redesignated it as the XJ79. A demonstrator engine with variable stators, the GOL-1590, was also authorized. Development of the new engine was supported by running the GOL-1590 demonstrator engine. The two other engines offered by GE, an advanced version of the existing J73 and a much larger design, known as the J77, were both cancelled. The GOL-1590 demonstrator engine ran for the first time on 16 December 1953. The first prototype of the production version, XJ79, ran on 8 June 1954.

The first flight of the XJ79 engine was on 20 May 1955 with the engine installed in a GE J47-powered North American B-45C Tornado (serial number 48-009). In flight the XJ79 was lowered from the bomb bay into the airstream for testing. The first flight after the 50-hour qualification test, required for a new engine that is the sole source of thrust for a flying testbed, was on 8 December 1955. The XJ79 powered the second pre-production Douglas F4D Skyray in place of its original Westinghouse J40 engine as part of the GE development and qualification program. The YF-104 was the next airplane to fly with the XJ79 followed by a re-engined Grumman F11F Tiger in a Navy-sponsored program to gain experience with the engine before the first flight of the F4H.

The *Military Characteristics No. 345 (SAB-53-A1) for the B-58 High-Altitude Bombardment System* was finally completed on 11 September 1953. While the characteristics remained similar to the August 1952 design, a major change required a 20,000-pound bombload, which was a 10,000-pound increase. The new characteristics also required the B-58 to carry high yield thermonuclear, biological, and chemical weapons as well as conventional high explosive bombs. This change meant a significant reduction in fuel load and loss of range. Wind tunnel tests continued to produce high-speed (supersonic) drag figures. The elevons and rudder were not inherently balanced and depended on the rigidity of their actuating systems to prevent flutter. Engine positions (and the Mach 2.1 speed) caused concerns. Finally, it was clear the pod was too big and had to be redesigned.

In October 1953, the Air Force authorized a proposal submitted by Convair in September to shorten the pod length from 89 feet to 30 feet and to separate it from the fuselage by a pylon. The nose gear on the pod was eliminated. In addition, the search radar was taken out of the pod and put in the nose of the aircraft. External fuel tanks were added on the wing tips to compensate for the fuel capacity lost due to the shorter pod.[24]

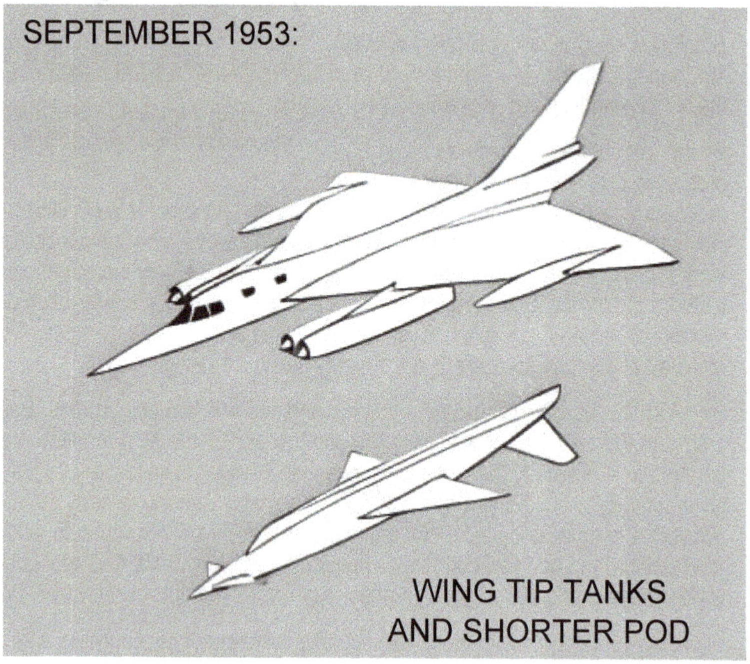

The revised design authorized in October 1953 shortened the pod from 89 feet to 30 feet and separated it from the fuselage by a pylon. External fuel tanks were also added on the wing tips. (Unknown)

B-58 Subsystems development proved more difficult than usual even though the Air Force expected it to take longer than airframe development. The B-58 was considered a complex, highly integrated, and mutually interdependent weapon system and the Air Force monitored every component's progress. They became especially concerned about the slow progress of the BNS and the defensive Fire Control System (FCS) used to aim and fire the tail gun, and the potential risks involved in developing these advanced systems planned for the B-58.

The Air Force consequently asked for studies to determine if the A-3A FCS could serve as a back-up for the planned active FCS being developed by Emerson. They also considered a modified IBM M-2 BNS instead of the planned AN/ASQ-42 BNS designed by

Sperry. A special committee was established to monitor BNS development. Bombing calculation was a serious issue and unlike anything previously encountered considering the planned B-58 supersonic speeds. For comparison, a B-17 flying at 25,000 feet required target identification at 11 miles for a 3-minute bomb run, whereas a B-58 flying at 40,000 feet and traveling at 450 knots required target identification at 25 miles for a 3-minute bomb run. Worse yet, a B-58 flying at 50,000 feet and traveling at Mach 2.0 required target identification at 66-70 miles out.

Configuration III was developed from 1953-54 to address some of the issues from Configuration II testing and incorporate the changes authorized in October 1953. Takeoff weight increased to 150,000 pounds. The navigator/bombardier and DSO positions were reversed, and the previously proposed jump seat was eliminated.

The Configuration III mock-up addressed some of the issues from Configuration II testing and incorporated the changes authorized in October 1953. (U.S. Air Force)

But not all changes were successful. Wind tunnel tests indicated that Convair's drag calculations were still wrong. These calculations were based on a previous agreement between Convair and the Air Force to move the engines from two split nacelles to two strut mounted "Siamese" nacelles. This change was intended to save weight, ease engine maintenance, and facilitate the retrofit of J57 engines with new J79 engines when they became available. However, the wind tunnel tests showed that this change induced excessive drag and the Air Force proposed that additional work be delayed while new tests were run.

On 23 June 1953, NACA released the results of the new wind tunnel testing. Previous testing had demonstrated that the Siamese configuration caused excessive drag over the split nacelle configuration at speeds from Mach 1.6 to 1.9 when the pod was attached even though there was little affect without the pod. The NACA test results confirmed that the split nacelle configuration was preferred. This required the return to split nacelles and more testing, which delayed the Configuration III mock-up inspection from May to September 1954. The Air Force authorized construction of airframe components including the front wing spar, secondary wing spars, and the chordwise bulkheads on 4 December 1953 to salvage as much of the schedule as possible.

The Air Force had invested $200 million in the B-58 by 1954 and still showed no tangible benefits. Consequently, the B-58 program was restructured from 3-4 March 1954 and approved by SECAF Harold E. Talbott on 30 April 1954. The restructured program emphasized R&D and delayed production. No change was made to the projected 1956 date for the first test aircraft, but the operational date of 1958 was no longer specified. Production aircraft quantity was reduced from 244 to 30 and these aircraft were referred to as "test vehicles." Long lead items (parts, training devices, support and test equipment) were cancelled. The first nine aircraft (rather than the first 18) would now have J57-P-15 engines and the remaining aircraft would have J79-GE-1 engines. The resulting Air Force procurement directive issued on 29 June 1954 freed about $190 million for 13 test aircraft, but no other procurement could be initiated until a final configuration was approved.

An Engineering Development Inspection of the Configuration III design held from 12-14 May 1954 was a complete fiasco. The inspection confirmed the poor results of the wind tunnel tests and, among other issues, SAC still insisted on side-by-side seating for the pilot and navigator/bombardier for improved crew coordination; the addition of an ECM pod to complement the pods already programmed; and the installation of an existing BNS, suitably modified, rather than launching into a development program for a new one. Also, on 5 May, ARDC had made the decision to use the GE 20mm T-171E-3 gatling gun in conjunction with the new bomber's active defense system. Several units by different manufacturers, covering various bores, firing rates, and firing velocities had been considered, but the GE weapon had proven undeniably superior.[25]

NACA researcher Richard T. Whitcomb, who had developed the Transonic Area Rule, was mystified by the drag problems encountered when aircraft designed using the rule were tested at supersonic speeds. However, by the summer of 1954 additional research by Robert Jones of the NACA Ames Aeronautical Laboratory demonstrated that the position and extent of fuselage narrowing was indicated by the aircraft's designed speed. Convair engineers finally accepted these findings, and they realigned the

Configuration III fuselage narrowing to comply with this modified Transonic Area Rule. Further wind tunnel tests confirmed that this change substantially reduced surface drag and a significant increase in supersonic performance could be expected.

Convair development of a new Configuration IV was completed in August 1954 and incorporated the Configuration III modified Transonic Area Rule changes. Configuration IV changes also included the elimination of the external wing tanks and a redesigned fuselage to add more room for fuel. All four engines were suspended on separate pylons (two on each wing). The tail area was extended to 160 square feet. Engineers determined that side-by-side seating, as SAC insisted, was not possible due to the required narrow fuselage design. Instead, the aircraft carried a three-man crew of pilot, navigator/bombardier, and DSO seated in tandem each in a separate crew compartment. Convair assured the Air Force that this configuration would meet the Military Characteristics but might require minor refinements. They were confident it was the best design supportable by the current state-of-the-art.

The Configuration IV design represented clear technical advances and promised significant supersonic capabilities. However, the production start had already slipped to 1958

The Configuration IV mock-up incorporated the modified Transonic Area Rule which substantially reduced surface drag and increased supersonic performance. (U.S. Air Force)

during the program restructuring in April 1954 and would slip once again up to seven months if the final design was not approved soon. By September 1954, the planned first flight had already slipped from January to June 1956, and then to 15 July 1956, and this date would slip on a day-to-day basis until Convair was authorized to proceed with tooling and parts manufacture. Unfortunately, the many differing opinions among Air Force leaders made approval uncertain. Many leaders doubted the new configuration would meet Convair expectations and refused to approve it. Many doubted that a Mach 2.0 bomber really made sense.

SAC leadership, including LeMay, continued to resist the B-58 aircraft based on their previous arguments from November 1952. They preferred larger aircraft with greater range and payload. LeMay said during a conference at ARDC on 21 December 1954, "after a thorough analysis of the B-58 program, it is concluded that the B-58 cannot be developed to encompass sufficient radius of action and defensive capability required for an acceptable Strategic Bombardment Weapons System." He followed this up in a 4 January 1955 memo to General Nathan F. Twining, Air Force Chief of Staff, that SAC neither wanted nor needed the B-58, since its range was only half that of the B-52. LeMay urged ARDC to develop a large supersonic intercontinental bomber to succeed the B-52.[26] As a final effort to kill the B-58, a mid-1954 study by Major General John P. McConnell, SAC Director of Plans, excluded the B-58 from the 51-wing bomber force proposed for 1958-65.

SAC leadership, including LeMay, preferred larger aircraft with greater range and payload. LeMay urged ARDC to develop a large supersonic intercontinental bomber to succeed the B-52 which ultimately led to the XB-70. (U.S. Air Force)

Letter Contract AF33(038)-21250 was modified again on 13 October 1954 authorizing Configuration IV development. But, by early 1955, the Air Staff began to reconsider SAC objections to the B-58 including LeMay's January memo. Consequently, SECAF Talbott appointed a B-58 Review Board on 2 February 1955. The board was chaired by Major General Clarence S. Irvine, AMC Deputy for Production. The objective was to determine if the B-58 should be "continued, modified, or cancelled." Major General Al Boyd, a member of the board, said even the Configuration IV design might not meet all requirements of the Military Characteristics, but he believed it should be continued even if not used as previously intended. He emphasized that the B-58 represented major technical advances and, therefore, entailed technical uncertainties and the risk of high costs. He said these would remain until the aircraft was flown and "we must accept such a risk sooner or later."

The advanced design of the B-58 delta wing bomber is clearly evident in this photograph of the B-58, B-52, and B-36 in-flight. (University of North Texas Libraries)

Lieutenant General Thomas S. Power, ARDC Commander and later LeMay's replacement at SAC, said the B-58 was the first opportunity to bridge the gap between subsonic and supersonic flight, a problem which demanded extensive knowledge of aerodynamic heating and aircraft materials. He agreed with Convair's chief engineer that the B-58 effort "would produce the earliest and most inexpensive integrated weapon system, as well as a very outstanding bomber. At worst, the B-58 would be superior to the existing B-47 medium range bomber." Power also thought the B-58 could fill Tactical Air Command (TAC) needs for a short-range attack bomber.

Major General Irvine agreed that the aircraft was "probably" worth the money already spent but after two years and $200 million "no tangible achievements could be claimed." However, if cancelled all money would be lost, whereas another $300 million could build the 13 aircraft already planned for the test program. The board submitted its recommendations to the Air Force Council and the SECAF on 10 March 1955. They said the restructured program approved in April 1954 should be continued with modification. Only 13 test aircraft should be ordered, and all should be equipped with J79 engines. A possible procurement of production aircraft for the operational inventory should be considered. All back-up systems efforts should be stopped to reduce unnecessary costs. The board also recommended that Convair should be asked to submit new configuration designs for a tactical bomber, a reconnaissance aircraft, and a long-range interceptor.

SECAF Talbott approved the development of 13 test aircraft only, with no production aircraft, on 2 June 1955. He directed ARDC to prepare a plan to utilize the 13 B-58 test aircraft and present it for final approval by 30 June.[27] Talbott insisted that the aircraft would be used for R&D only and the Air Force would no longer consider producing a high-altitude strategic bomber and reconnaissance system within the B-58 program. The R&D program would help the Air Force determine a suitable role for the aircraft since SAC did not want it as a strategic bomber.[28] It would also allow engineers to learn about aerodynamic problems at high-altitudes and supersonic speeds. Finally, it would allow for testing of subsystems for use on future weapon systems, a nuclear-powered aircraft being studied by Boeing, and a tactical bomber identified as Weapon System 302A. Talbott also required cost reductions and wanted the first flight to occur by November 1956.

AMC quickly responded to most of Talbott's direction including cost reductions. They had already been studying costs and estimated that $554 million was needed for 13 aircraft, pods, engines, GFE, and Convair's fee. This estimate included identifying $50 million in reductions for items not needed due to the R&D-only role. It also included eliminating non-essential B-58 subsystems, while continuing development of systems that could be used for other programs. Convair eagerly matched AMC cost estimates and also provided estimates for up to 500 production aircraft, believing that a successful test program would eventually lead to a production contract. Consequently, AMC authorized Convair to continue engineering, tool fabrication, airframe parts procurement, etc. in mid-June.

Talbott's directives were short-lived, however. He retired on 1 August 1955 and the SECAF position was left vacant for two weeks. General Putt, who was now the Air Force DCS for Development, pushed the program he had helped to initiate believing it could be "a useful SAC tool." General Irvine, now DCS for Materiel, and others on the Air Force Council agreed. General LeMay was still not convinced. Nevertheless, General Twining approved the directive after extensive debate and the B-58 once again became a production candidate. Twining's approval was apparently upheld by Donald A. Quarles who succeeded Talbott as SECAF in mid-August. The directive issued on 22 August 1955, apparently worded to take LeMay's objection into account, called for a wing of B-58 aircraft without mentioning SAC or any specific command assignment.

Letter Contract AF33(038)-21250 was finally definitized on 23 November 1955 by Definitive Supplemental Agreement No. 26. The contract ordered the Convair B-58 supersonic bomber into initial production but covered only the 13 R&D aircraft (55-0660 to 55-0672) and 31 pods (12 MA-1, 16 MB-1, 2 MC-1, and 1 MD-1), spares, ground support equipment, and test support equipment authorized for FY55 as required by the previous letter contract modifications. The contract required that the B-58 program be carried out under the Air Force's new Weapons System Development policy. It added $340 million, bringing total contract value to $540 million. It also included a Cost-Plus Incentive Fee (CPIF) which paid Convair an additional fee based on exceeding technical performance requirements, weight control, and adherence to cost and schedule.

One of the first 13 B-58 R&D aircraft (55-0664) on the ramp at Convair on 16 January 1958. (University of North Texas Libraries)

A new Letter Contract AF33(600)-32841 was awarded just a few months later on 25-May 1956. It provided another $13.6 million for long-lead items and to sustain the minimum production rate through October 1956. It also provided an option to procure the remaining 17 R&D aircraft. If exercised it would allow procurement of 17 aircraft, 17 powered bomb pods, 12 free fall bomb pods, 3 photo pods, and 3 ferret (electromagnetic data) pods. However, the Air Force initially opted to delay the decision until the fall of 1956. The contract also included options that would support the 22 August 1955 SECAF decision to buy a wing of B-58 aircraft based on successful R&D test results.

The B-58 first flew in November 1956 and quickly began racking up successful test flights. It had flown up to 35,000 feet and at speeds up to Mach 1.6 clearly demonstrating its supersonic capability. These successes, although still early in the test program, showed that the B-58 was a viable weapons system and positively influenced Air Force leadership. In early 1957, General Otto P. Weyland, TAC Commander, requested a minimum of two B-58 wings supporting the short-range attack mission. But the B-58 was being developed as a SAC aircraft, and it would require several time-consuming and costly changes to meet TAC mission requirements.

SAC, including LeMay, had initially resisted the B-58 since they wanted a larger aircraft and were counting on an advanced bomber development program that would eventually be designated the XB-70. However, XB-70 program delays caused SAC to become more involved with B-58 development, and they began gearing up to support the test program in late 1956. They now thought of the B-58 as a possible replacement for some or all B-47 aircraft due to its small size and supersonic speeds. With one refueling 2,500 miles from its takeoff in the "logistics zone", it appeared to have sufficient range reserve for most targets. The aircraft, SAC believed, would be dependent upon forward bases for deep penetration, but it could be operated as an intercontinental bomber on a limited basis if advance bases were denied.[29]

By spring 1957, the B-58 had achieved Mach 2+ speeds and was established as the world's fastest jet bomber. This success caused a decrease in the B-52 production rate due to budget decisions affecting SAC and other Air Force aircraft programs. LeMay quickly dropped any newfound support for the B-58 and reasserted that no B-58 aircraft were needed for SAC. He believed that the B-52G equipped with new low-level penetration aids would be superior to the B-58A or any improved version. He also believed that the XB-70 would be an improvement over the B-52G, unlike the B-58, due to increased range and speed. He therefore recommended that the B-58 be limited to a test program with no funding authorized for production or model improvements.

But the Air Staff again disagreed with LeMay and insisted that SAC needed a supersonic bomber as soon as possible. They pointed out that the decision had already been made to buy a limited quantity of production B-58A aircraft for SAC. Furthermore, they said B-58 development was important to protect the current technological lead over the Soviets and the money invested in the B-58 would not affect the XB-70. Finally, they expected that B-58 reliability and capabilities would improve through normal development growth.

The option for 17 additional R&D test aircraft (58-1007 to 58-1023) included in Letter Contract AF33(600)-32841 was finally exercised in mid-1957 for FY58 procurement to supplement the 13 aircraft already being used. The original 13 aircraft supported a series of corporate service tests at the Convair plant in Fort Worth, TX, referred to in Air Force historical aircraft records cards as Convair Vultee Corporation Fort Worth (CVCFW), and the Air Force Flight Test Center (AFFTC) at Edwards AFB, CA.[30] Most of the additional 17 aircraft were assigned to the 6592nd TS and used for performance and operational tests at Carswell AFB, TX.[31]

A B-58 R&D aircraft (55-0667) undergoing corporate service testing at CVCFW. (University of North Texas Libraries)

In 1958, the B-58 again faced near cancellation. It became clear that the B-58 could only obtain a combat radius of 4,500 miles with one refueling. The limited testing performed so far had uncovered several issues which required costly configuration changes recommended on 21 February by an 85-man team from Convair, ARDC, AMC, and SAC. The recommended changes included a Two-Component pod (later known as a TC pod); the use of a single-side band High Frequency (HF) radio; the use of an emergency Ultra-High Frequency (UHF) radio; the development and use of encapsulated ejection seats (in response to several B-58 accidents which should have been survivable using ejection); the use of Tactical Air Navigation (TACAN); the incorporation of a "deadman switch"[32]; and the deletion of the ALD-4 ferret system requirement.[33]

General Power, who replaced LeMay as CINCSAC in July 1957, told the Air Staff that the problems were exaggerated, and he believed a mixed B-52 and B-58 force was the best way to replace the B-47. General Thomas D. White, Air Force Chief of Staff, agreed but cautioned that the FY59 quantities should be reduced by one-third. The money saved would pay for the most important changes and the inevitable cost increases. General LeMay, now Air Force Vice Chief of Staff, was the only member of the Air Force Council to still oppose the B-58. He believed the refueling costs alone did not favor the B-58 since it would take one tanker to refuel one B-58, whereas two tankers could service three B-52s.

A new Letter Contract AF33(600)-36200, which had been issued on 1 November 1957, contemplated procurement of 47 production B-58A aircraft using FY59 funding, bringing the total to 77 aircraft including the 30 R&D aircraft. However, these 47 aircraft were reduced to 36 B-58A aircraft (59-2428 to 59-2463) when the contract was definitized on 26 September 1958 bringing the fleet total to 66 aircraft.

In addition to reducing the planned FY59 production order, the Air Force implemented several other cost and capability changes by the end of December 1958 as previously recommended by the 85-man team. The photo reconnaissance requirement was deleted. The ME-1 pods and ground photo processing equipment, as well as 45 ALD-4 ferret pods were all cancelled. The MB-1C free fall bomb pod was replaced with a TC bomb and fuel pod to extend range. This arrangement allowed dropping the lower component after the fuel was used and the upper bomb and fuel pod could be dropped on target to lighten the return flight load. Improved communications systems were added including a single side-band HF and emergency UHF radios. Encapsulated crew ejection seats, TACAN electronics, and various minor improvements were also added.

The B-58 program came under scrutiny, yet again, during 1959-60. A RAND study requested by the Air Staff supported LeMay's contention that the B-52 was superior to the B-58. The B-52 could carry heavier loads and had a longer range. Refueling could extend B-58 range but would be "unreliable and expensive." RAND also considered each aircraft's ability to penetrate enemy defenses but concluded it was unimportant because future enemy defenses would be so sophisticated that bomber losses would be high regardless of speed. RAND concluded the most economical alternative would be to equip B-47 aircraft with improved engines. But the Air Force rejected this alternative because, although the B-47 would be somewhat improved, it would still be mostly obsolete. They asked RAND for a second review, but it produced the same results.

Still, the RAND study fueled growing dissatisfaction with the B-58. Correction of deficiencies was slow, and many components were in short supply meaning early aircraft would be fielded without some critical systems including HF radio and Identification Friend or Foe (IFF) equipment. Perhaps the most critical, the newly required low-level capability modifications would be extensive and costly. Some SAC officials now believed two B-58 wings would be sufficient due to the greater tanker support required than for the B-52. Other SAC officials wanted more, banking on improvements made in the planned B-58B. In May 1959, General White endorsed the B-52, XB-70, and nuclear aircraft with no mention of the B-58. However, the Air Force issued a letter contract in the same month for long lead items to support production of 39 more B-58A Hustlers in FY60 bringing the total to 105 aircraft.

Despite these issues, Air Force planning, which began on 11 June 1959, included 185 B-models bringing the total fleet to 290 aircraft. This would allow equipage of five wings with the first wing becoming operational in November 1960. The B-58B was expected to provide low-level capability, increased range, speed, altitude, and external stores including free fall bomb pods, fuel tanks, and air-to-surface missiles. Not surprisingly, LeMay continued to object, stating the aircraft was too expensive and the low-level equipment would be too complex.

But all this progress was fleeting. By 7 July 1959, the Air Staff eliminated the B-58B program and came close to cancelling the B-58A. Letter Contract AF33(600)-38975 reduced 60 B-58A aircraft planned for FY60 production; first to 32 and then to 20 aircraft. The 39 B-58A aircraft planned for FY60 procurement were also cancelled. General Power tried to justify the 290-aircraft program, but on 14 July he was informed by SECAF James H. Douglas that the B-58B was a dead issue due to the lack of funding. Only previously funded development and potential for near term availability stopped complete cancellation of the B-58A. Program slips, and lowering priority, were weakening the B-58A production case; originally designed for 1958-65 targets the aircraft would not be available before 1962. By December 1959, SAC plans called for only 148 aircraft.

A subcommittee of the U.S. Senate Armed Services Committee, called the U.S. Senate Preparedness Investigating Subcommittee and headed by Texas Senator Lyndon B. Johnson, issued a report in July 1960 after a thorough review of the entire B-58 program. The report covered B-58 capabilities, history, production status, flight test program, growth potential, and provided recommendations to the Secretary of Defense (SECDEF). It emphasized the continuing need for manned bombers in the nation's deterrent force. It also cited several significant milestones achieved by the B-58 program including flights of 12-18 hours durations; a Seattle to Fort Worth flight completed in 80 minutes; four minutes to gear-up scramble; pod drops at Mach 2.0; and launch of Air Launched Ballistic Missiles (ALBM) at speeds up to Mach 2.0.

The report remarked that the Mach 2.0 capable B-58 Hustler had "in one giant step achieved a greater speed increase over the previous fastest strategic bomber than was reached in the preceding 50 years." It concluded, "The B-58 is the only airplane presently operational which could fulfill (the requirement for a follow-on manned weapon systems) in the near future – and lack of funds should not be utilized as a limiting factor in providing our Air Force with a technologically modern inventory of manned bombers."

Although the report sang the B-58's praises, it was not enough to keep the budget hackers away. A July 1960 Letter Contract AF33(600)-41891 included 30 B-58A aircraft and 96 BLU-2/B TC pods planned for FY61 procurement, but they were subject to cancellation. Ultimately, these would be purchased but all planned FY62 or later procurements were cancelled. The total B-58 procurement was set at 116 aircraft, only half of the planned fleet, at a cost of $3.2 billion (FY61 dollars). This meant SAC would have to be satisfied with only two B-58 wings.

Contract	Date	Funding	R&D	B-58A	B-58B	Total
AF33(038)-21250	December 1955	FY55	13			13
AF33(600)-32841	25 May 1956	FY58	17			30
AF33(600)-36200	1 November 1957	FY59		47		77
AF33(600)-36200	26 September 1958	FY59		(11)		66
Unknown	Unknown	FY60		39		105
Unknown	11 June 1959	TBD			185	290
Unknown	7 July 1959	TBD			(185)	105
AF33(600)-38975	7 July 1959	FY60		60		165
AF33(600)-38975	Unknown	FY60		(28)		137
AF33(600)-38975	Unknown	FY60		(12)		125
Unknown	Unknown	FY60		(39)		86
AF33(600)-41891	July 1960	FY61		30		116
AF33(600)-41891	July 1960	FY62		32		148
AF33(600)-41891	December 1960	FY62		(32)		116
Total			**30**	**86**	**0**	**116**

The Air Force ultimately purchased 116 B-58 aircraft for $3.2 billion (FY61 dollars) under various contract actions.

Despite some drawbacks, the aircraft introduced several technical advances. Manufacturing processes departed from riveted metal and used composite structural and adhesive methods instead. This significantly reduced aircraft drag and improved supersonic capabilities. The design addressed technical uncertainties for flying aircraft at supersonic speed and took developmental risks to advance aerospace technology. The aircraft set several world records and won all major aviation awards. It was also used as the basis for a new Convair 52-seat supersonic transport in mid-1961. However, this concept never proceeded beyond design stage.

The B-58 became the only operational supersonic bomber despite some competition. Two of the competitors were derivatives of tactical aircraft including the French Mirage

IV which as logical progression of the Mirage interceptor and the FB-111A which was bomber version of the TAC fighter. Two, like the B-58, were developed as bombers from the beginning. The XB-70 was designed as a Mach 3 bomber and was selected by the Air Force for development in December 1957. However, it was cancelled by President Kennedy in 1961 and only two XB-70 (62-0001, 62-0207) aircraft were built. One aircraft (62-0207) was destroyed after a collision with an F-104 chase plane in June 1966. The other (62-0001) was placed on display at the National Museum of the United States Air Force in February 1969. The Rockwell International (formerly North American Aviation) B-1A, which was designed as a Mach 2.2 bomber, was cancelled in 1977 by President Carter in favor of equipping the B-52G and H models with stand-off missiles. Four prototypes were built and flight tested through April 1981. The aircraft was subsequently revived as the B-1B under President Reagan, but its design was altered to focus on low-level, high-speed penetration and its maximum speed was limited to Mach 1.2.

Models and Variants

R&D Aircraft

YRB-58A. The B-58 Hustler test regime, conducted under the new Weapons System Development program approach, deviated significantly from the approach used with previous aircraft acquisition programs. Previous programs typically developed an X-prototype (experimental) used for testing conducted by contractor flight crews to verify the basic design including aerodynamics, flight characteristics, handling, and system functions. This was followed by a Y-prototype (production) that was used for further developmental and operational testing of the aircraft and its systems conducted by a mixture of contractor and Air Force flight test crews. Only then were production aircraft given to the Air Force operational command, such as SAC, for operational testing.

Convair YRB-58A (55-0660) rotates during high-speed taxi test. (U.S. Air Force)

This approach often left Air Force senior leaders, including LeMay, wishing for more testing to flush out and correct all problems before starting production. Unfortunately, many problems were subsequently discovered after initial fielding of the production aircraft which led to modifications during and post-production. These modifications often delayed the aircraft's Initial Operational Capability (IOC) and Full Operational Capability (FOC) dates.

This time, LeMay insisted on testing with more aircraft before starting production. Therefore, the B-58 program office implemented the testing approach envisioned by the

Cook-Craigie production policy. They planned to procure 30 aircraft for R&D testing while preparations were made for full scale production of a version incorporating all test-dictated changes.[1] These aircraft would be built under two contracts with the first delivering 13 aircraft and the second contract delivering 17 aircraft. All changes made during R&D testing would be incorporated into the final production configuration and the first 30 aircraft would be retrofitted to the approved configuration. This plan was expected to eliminate faults in the initial design before starting final production. It was easily one of the most rigorous test programs ever undertaken for a new aircraft. However, it was expected to cut as much as 12 months off the time formerly required to take a new aircraft through test and training programs.

Convair YRB-58A (55-0660) was the first of 30 R&D aircraft. (U.S. Air Force)

The B-58 program did not include an X-prototype despite some references stating that the first two aircraft were designated XB-58A. They were not. Other references show these aircraft and the next 11 as YB-58A. They also show the final 17 aircraft as YB-58A or RB-58A. Still other references show all 30 aircraft as YB/RB-58A. This probably closest to the truth. Despite the lack of definitive documentation, it appears likely that the first 13 aircraft were ordered as YB-58A. The next 17 aircraft were likely ordered as YB-58A also, but they were later redesignated as YB/RB-58A with the expectation they would be modified to carry reconnaissance pods. Eventually, the YB/RB-58A designation was apparently applied to all 30 of the test aircraft since all could carry reconnaissance pods. Furthermore, according to the initial flight manual dated 28 August 1959, production aircraft were also initially referred to as B/RB-58A.

According to official aircraft record cards obtained from the Air Force Historical Research Agency, all 30 aircraft were designated as YRB-58A at CVCFW before delivery. It is likely that YB/RB-58A was shortened to YRB-58A to fit in the seven-character limit of the cards. The records also show that AMC incorrectly recorded the first 14 aircraft (55-0660 to 55-0672, 58-1007) as B-58A, but these were corrected to YRB-58A on 22 December 1958. The YRB-58A designation is reflected in other references including some aircraft crash reports. Therefore, this book uses the initial designation of YRB-58A for the first 30 aircraft to remain consistent with Air Force historical records.

The test program was affected by Air Force insistence that requirements reflect technological developments (USAF engineers had previously asked for more realistic military characteristics and advocated for state-of-the-art design compromises). It was conducted in three phases, or categories, designed to address the highly complex integrated systems used in the B-58. Category I, accomplished by Convair, included subsystem development test and evaluation, as well as basic flying characteristics of the aircraft. Category II, conducted by Convair and the Air Force, focused on systems development tests and evaluation. Category III was conducted by the Air Force and resulted in the development of techniques and procedures for using the B-58 as a weapon system.

Production of the first YRB-58A (55-0660) was completed in late August 1956, and it rolled off the factory floor on 4 September. The first engine run-up occurred on 1 October and taxi tests began on 29 October. Category I testing, which accumulated about 3,000 hours of flying time, began in November 1956 at CVCFW with some initial difficulties. A tire blew during a high-speed taxi test on 5 November, and on 7 November the drag chute was deployed for the first time. Unfortunately, smoke began pouring from the landing gear when the brakes were applied. These issues were quickly fixed. The first flight occurred on 11 November 1956 (Veteran's Day and a Sunday), which met the date specified in the 2 June 1955 SECAF approval, although it was 15 days behind Convair's proposed schedule. The aircraft was packed with test equipment and there was very little operational equipment installed. It carried a light fuel load and no pod was attached. The flight lasted only 38 minutes. This was also the first public unveiling of the aircraft, which had previously been shrouded under tight security.

The first flight was piloted by Convair Chief Test Pilot B.A. Erickson. The crew of three also included John. D. McEachern as the Systems Specialist and Charles P. Harrison as the Flight Test Engineer. Erickson described the flight as "outstandingly successful and very satisfying." This flight was followed by a second flight on 14 November and a third flight on 4 December. All flights took off from CVCFW using the same aircraft, YRB-58A (55-0660), and were considered successful. The second flight lasted one hour and reached Mach 0.94 at 35,000 feet altitude. The aircraft was the first B-58 to achieve supersonic speed when it flew Mach 1.17 at 35,000 feet for nine minutes during the seventh test flight on 30 December 1956. This was topped on the same day when the aircraft reached Mach 1.31 at 35,000 feet during the eighth test flight. On 22 January 1957, it made the first flight with an Air Force pilot and reached Mach 1.35. It also became the first aircraft to fly at night. YRB-58A (55-0661) made its first flight on 16 February 1957 with an MC-1B pod attached reaching Mach 1.15, just two days after the last B-47 rolled out of the plant. These flights marked the beginning of the R&D program, and these successful results would positively influence later Air Force decisions.

The B-58 first flew on 11 November 1956 and quickly showed that it was a viable weapons system with Mach 2+ supersonic capability. (U.S. Air Force)

Category I testing of the first three YRB-58A aircraft was successful overall. By the end of 1957, the three aircraft combined had attained Mach 2.11 at all altitudes over 50,000 feet. They made two successful pod drops at Mach 2.0 from 42,000 ft and proved the ability to maintain Mach 1.15 for 91 minutes. They had also shown the ability to "zoom" without a pod from Mach 2.0 at 50,000 feet to Mach 1.13 at 68,000 feet.

Convair YRB-58A (55-0660) touches down after its first flight. (U.S. Air Force)

On 24 July 1957, Convair test pilot B.A. Erickson landed a YRB-58A at Wright-Patterson AFB, OH marking the first visit of a B-58 away from the CVCFW plant. The primary purpose of the routine flight test was to obtain radar data and stage the aircraft for take-off at increased gross weight. The aircraft logged the highest takeoff weight to date of 131,000 pounds. It attracted significant interest among Air Force officers and civilians at the base, which was noteworthy since they were accustomed to seeing every new type of aircraft built and they were nonetheless eager to see the B-58. The elapsed time of the 775-mile flight was 66 minutes at an average ground speed of 811 mph. True airspeed varied from Mach 1.4 to 1.56 at 43,000 feet logging 46 minutes at supersonic speed. The aircraft performed beautifully according to Erickson. When asked by a reporter if supersonic flight was routine, Erickson replied "It is with this baby."

One flight on 3 September 1957 also carried General Power, CINCSAC, piloted by test pilot B.A. Erickson. General Power stated that it was "a very fine, very good flight." He flew faster than the speed of sound but commented that it was "not as thrilling as it sounds. As a matter of fact, if you didn't look at the instruments, you wouldn't even know you were going faster than sound. It's a very smooth ride." He later stated in a press conference that the B-58 was not a direct replacement for any particular aircraft, but a "bombing platform" whose performance and reliability would be established during the flight test program.

The first YRB-58A (55-0660) during maintenance in Run Station No. 2 (back view) at CVCFW on 27 December 1956. (University of North Texas Libraries)

The flight with General Power was followed on 18 September by a two-hour flight carrying Lieutenant General C.S. Irvine, DCS for Materiel. In response to a reporter's question about flying in the world's fastest bomber, he said "The flight was rather boring because nothing unusual happened." He agreed with Power that it was impossible to tell how fast the plane was traveling without monitoring the instruments. However, he also stated "we certainly are pleased with the performance of this airplane and I'm quite sure it's going to fill an important place in a family of weapons we have in the Air Force."

Despite these successes, several issues were uncovered. The aircraft were initially produced using prototype YJ79-GE-1 engines pending J79-GE-5 certification. The YJ79-GE-1 engine was an early test version of the J79 and nominally rated at 8,900 pounds normal thrust, 9,300 pounds military thrust, and 14,350 pounds thrust each with afterburner. It was basically an experimental engine and was not capable of sustained operations with any regularity. Mean time between engine overhauls was very limited and numerous teething problems were encountered. It was, however, the first Mach 2-capable production turbojet in its class.

The first YRB-58A (55-0660) during maintenance in Run Station No. 2 at CVCFW on 27 December 1956. (University of North Texas Libraries)

While J79-GE-1 engine availability negated the need to use the P&W J57 engines for testing as originally contemplated, the prototype engines had several faults. These engines experienced afterburner issues which caused yawing at supersonic speeds. Vibration issues caused acoustic and sonic fatigue in the aft fuselage which threatened to cause testing restrictions if not resolved. Vibration also caused cracks along the rivet lines in the forward fuselage after only 50 flight hours, which were expected to worsen with the more powerful J79-GE-5. These engine vibrations could also affect the electronic equipment installed in the aft fuselage and the aft portion of the pods.

Convair YRB-58A (55-0661) sits on the ramp at CVCFW and clearly shows its equipped YJ79-GE-1 engines. (U.S. Air Force)

The test aircraft also experienced fuel system malfunctions which caused fuel to slosh around during acceleration/deceleration and impaired flight stability. This initially delayed the flight test program until Convair implemented several modifications on YRB-58A (55-0660) including incorporation of additional fuel warning lights and an inclinometer; segregation of the rear fuselage fuel tank with provisions for pumping directly to the main fuel manifold during deceleration from high supersonic speeds; and Isolation of fuel in an extreme forward area plus automatic fuel transfer to maintain the desired Center of Gravity (CG) to assure engine fuel feed and accurate fuel quantity gauging during supersonic acceleration and deceleration.[2]

Inadequate heat dissipation after braking caused frequent tire failures when landing at high gross weights and during high-speed taxi runs. BNS development slippages threatened delays in initial equipment deliveries and the resulting test program schedule. Finally, the ejection seats were unsafe at high speeds due to insufficient thrust for launching the seat clear of the aircraft. Convair tested a rocket-type catapult seat and other types of seats, but they also presented problems. The Air Force and Convair preferred a capsule-type escape system (under development by both Martin and Goodyear) but "time was of the essence."[3]

One engineless B-58 airframe (never assigned a serial number but built between 55-0663 and 55-0664 and referred to as B-58 No. 4A within Convair) was allocated to long-term fatigue testing. The airframe was delivered to the Wright ADC Structures Test Laboratory at Wright-Patterson AFB, OH on 12 March 1957 by adapting a B-36F (49-2677) as a transport and carrying the airframe underneath its fuselage in much the same manner as had been intended in the original parasite program. To do this, the B-36's inboard propellers had to be removed, and a temporary shackle system was attached to the bomb hoist mechanism. The ground clearance for the suspended B-58 airframe was only 22 inches. With the B-58 in place it was impossible to retract the B-36 landing gear, so the flight was made with the gear down. The photographs released of the B-36/B-58 airframe combination were misunderstood by some who imagined that the B-36 had launched the B-58 in mid-air![4] The airframe was tested to destruction in February 1962 during a 5-year fatigue test program. It was subjected to a 115 percent load followed by two 135 percent loads simulating a pull-up with maximum gross weight and multiple weapons before failure.

B-36F (49-2677) transporting B-58 No. 4A fuselage to Wright ADC for long-term fatigue testing. (National Archives)

YRB-58A (55-0670) was sent to Eglin AFB, FL from 8 July 1958 through September for two months of climatic hangar evaluation under temperatures ranging from desert to arctic conditions. The aircraft was subjected to temperatures as low as 65 degrees below and as high as 165 degrees above zero. It was later flown to Eielson AFB, AK in

December 1958 for actual service testing under Operation RAW DEAL. The aircraft made the routine 2,870-mile flight in five hours. The aircraft, crew, and equipment were subjected to sub-zero temperatures down to minus 60 degrees to test its operational suitability in Arctic weather. The operation included 57 Air Force personnel and 40 civilians who remained deployed and operated in Alaska until 15 March 1959. The aircraft was serviced and maintained by Air Force maintenance personnel, with assistance from Convair if needed, during the deployment.

On 18 January 1959, YRB-58A (55-0666) made a flight to check acceleration performance. Beginning at 30,000 feet altitude and a weight of 130,000 pounds, the aircraft accelerated to Mach 2.0 and 36,000 feet in seven and one-half minutes.

Convair YRB-58A aircraft are worked on by maintenance crews in the Experimental Building at CVCFW on 28 October 1957. (University of North Texas Libraries)

YRB-58A (55-0662) became the first R&D test aircraft to complete Category I testing on 25 April 1959. By the end of May, Category I testing had proven quite successful and had generated the following data: on a classic SAC strike mission, the B-58A could strike 87 percent of the targets in the Soviet Union; the B-58 and KC-135A were totally compatible from an inflight refueling standpoint; the B-58 had demonstrated its ability to fire air-launched ballistic missiles; the B-58 had successfully penetrated U.S. air defenses three times without using its onboard ECM capability; the aircraft had demonstrated confusion and track-breaking ECM; the BNS had become more reliable and maintainable, and had demonstrated specified performance; the aircraft had demonstrated its ability to carry multiple stores, the aircraft had flown low altitude dash missions at Mach 0.9; and the aircraft had flown supersonically at Mach 2.0.[5]

Category II testing was originally planned to occur between March 1959 and June 1960 at Carswell AFB, TX. However, it got an early start on 15 February 1958 when ARDC assumed responsibility for some of the tests initially intended for Category I including pod drops, aerial refueling, and a few special tests. Category II testing was conducted at the AFFTC and Carswell due to its proximity to the Convair plant, which was expected to reduce costs for logistical training and Convair technical support. Testing was consolidated under the B-58 Joint Task Force (JTF), the first such JTF in the history of the Air Force and industry, headed by the ARDC B-58 Weapon Systems Office with Convair and SAC participation. Consolidated testing was expected to identify and solve problems quickly thus getting the aircraft through the test phase and into operation in the shortest possible time.

ARDC activated the 6592nd Test Squadron (TS) at Carswell on 1 March 1958 under the command of Colonel David M. Jones to support Category II testing and coordinate all B-58 JTF activities. SAC activated the 3958th Operational Evaluation and Training Squadron (OETS) at Carswell on the same day. This unit became an important member of the B-58 JTF and was responsible for proper combat crew training development. SAC participation in the test program provided valuable experience toward the formation of B-58 training and operational units. The 3958th selected and trained B-58 flight crews and developed a cadre of instructors for combat crew training classes, along with a cadre of maintenance personnel and instructors. It also developed Standard Operating Procedures (SOP) for future B-58 wings.

First Convair YRB-58A (55-0661) delivered to the 6592nd TS at Carswell AFB, TX on 29 August 1958. (U.S. Air Force)

The first delivery of a YRB-58A (55-0665) to the Air Force occurred on 14 February 1958. This aircraft was assigned to the AFFTC at Edwards AFB, CA for Category II testing under ARDC. The first delivery of a YRB-58A (55-0661) to the 6592nd TS occurred on 29 August 1958 at Carswell AFB, TX. It landed after a two-hour test flight on the Carswell runway, which was shared with the CVCFW factory where the aircraft was manufactured and initially tested. The aircraft was piloted by Air Force test pilot Major Leon Amos, who was the fourth Air Force pilot to be qualified in the B-58. Colonel David M. Jones, 6592nd TS Commander, officially accepted the aircraft for the Air Force.

Operation SEVEN-UP began in February 1958 when YRB-58A (55-0666) underwent maintenance and was dedicated as a special demonstration aircraft. Operation SEVEN-UP was considered even more important than the first B-58 flight because it was the chance to show how the aircraft would perform as an Air Force weapon system. The aircraft was of the latest configuration and was used for a series of mock military

YRB-58A (55-0666) aircraft undergoing maintenance as part of Operation SEVEN-UP. (University of North Texas Libraries)

missions of varying ranges and speeds. The first test flight was in April 1958 as the aircraft equipped with YJ79-GE-5 engines went into flight operations. It was the first flight of a series leading to three demonstration flights of the Hustler's military capabilities and range. The three demonstration flights occurred between June 27, 1958, and March 17, 1959. The first flight was flown at subsonic speeds on an outbound leg of 2,320 miles and inbound leg of 2,372 miles with a pod drop at the Holloman AFB test range in between. The total flight time was 8 hours and 55 minutes and the cruise speed varied between Mach 0.92 and Mach 0.93. The takeoff weight was 161,000 pounds.[6] All three flights resulted in pod drops over the Holloman range.

On 4 February 1958, YRB-58A (55-0660) began a 10-day project called Operation WHIRLWIND to determine sortie rate capabilities and pre-flight and post-flight maintenance procedures. It flew 11 missions for 10 consecutive working days totaling 37 flying hours and flew throughout its operating range to demonstrate the Hustler's durability. One flight was nearly six hours at speeds greater than any other known operational bomber. It flew at Mach 1.2 for 1 hour and 31 minutes on another flight, which was probably the longest sustained supersonic flight at that time. Finally, it completed three 360-degree rolls flying Mach 1.3 at 25,000 feet.

The first YJ79-GE-5 engines were finally retrofitted on YRB-58A (55-0666) in 1958, after arriving at CVCFW on 27 September 1957. These new engines were more powerful and much more reliable than the J79-GE-1. Each engine produced 9,370 pounds of normal thrust, 10,000 pounds military thrust, and 15,600 pounds of thrust with afterburner. The aircraft first flew with the new engines in April 1958 as part of its Operation SEVEN-UP modifications. It then flew for 32 minutes at sustained Mach 2.0 on 8 November 1958 allowing engineers to record the structural effects of sustained atmospheric friction on the bomber's honeycomb skin panels. This profile also proved that the Hustler could sustain maximum cruise speeds long enough to outrun contemporary Soviet interceptors and surface-to-air missile batteries over vast distances. YRB-58A (55-0666) ultimately completed 330 hours of accelerated service test under Category II at the AFFTC (and another 170 hours by SAC crews under Category III).

A YRB-58A flew over the nation's capital in May 1958 to celebrate Armed Forces Day. It was also on display that month at several Air Force bases for the first time including Andrews AFB, MD, Carswell AFB, TX, Kirtland AFB, NM, and Kelly AFB, TX. The aircraft also performed a flyby at the Air Force's annual firepower demonstration at Eglin AFB, FL.

Refueling tests with a KC-135 tanker were the key to greatly extending B-58 operational range. YRB-58A (55-0661) refueled from a KC-135 tanker for the first time on 11 June 1958. The flight took place at 33,000 feet and 10,000 pounds of JP-4 were successfully transferred. These tests were followed by another series on 23 July, when the B-58 was refueled from a KC-135A at an altitude of 30,000 feet and a gross weight of 156,000 pounds. The refueling trials were considered very successful and it was determined that the B-58 was totally compatible with the KC-135A. A final test on 8 August 1958 involved a 3-hour and 5-minute mission (45 minutes being at supersonic speeds) in which the B-58 was refueled a total of four times (during six connections) and received some 62,000 pounds of fuel.[7]

Convair YRB-58A (55-0660) flew faster than the speed of sound for more than an hour and a half in 1958 – probably the longest sustained supersonic flight at that time. (U.S. Air Force)

A YRB-58A (55-0660) flown by an Air Force crew performed a series of sonic boom tests over Wallops Island, VA in July 1959 to evaluate the effects of sonic booms on the civilian population. The Air Force had previously considered the effects negligible but increased complaints of damage due to the increase in sonic booms generated by the B-58 caused concern. These flights concluded that sonic booms caused a serious problem and that further testing was needed. Follow-on tests using the YRB-58A and other supersonic aircraft resulted in the elimination of supersonic missions over land except in a few restricted areas.

A YRB-58A (58-1015) flew the first sustained Mach 2.0 flight from Seatle, WA to Carswell AFB, TX at nearly 1,320 mph in 70 minutes on 15 October 1959. On 15 December 1959, the first production B-58A (59-2428) flew 1,680 miles in 80 minutes and set a new speed record at an average speed of 1,213.16 mph.

A YRB-58A (55-0671) named Mary Ann made a routine flight of 18 hours and 10 minutes from 22-23 March 1960 at 620 mph average to mark a new B-58 endurance record. The aircraft took off from Carswell AFB, TX at 6:48 pm on 22 March and returned at 12:49 pm on 23 March. The aircraft was flown by a SAC crew assigned to the B-58 JTF at Carswell including Lieutenant Colonel Leonard M. Legge (pilot), Captain Andrew Rose (navigator/bombardier), and Captain Ray R. Wagener (DSO). The aircraft was refueled twice by a KC-135 during the roughly 11,000-mile flight. This is believed to be the longest lasting single flight ever flown by a B-58. The previous longest B-58 flight was 13 hours and 10 minutes.

This flight was part of accelerated service tests of the new J79-GE-5A engines. It was the sixth test in 15 days of testing in which the aircraft logged 67 hours and 15 minutes of flight time. Overall, the J79-GE-5A endurance test lasted 58 days and logged 179 flight hours[8] from 9 March to 15 May 1960, including 67 hours in March, 51 hours in April, and 57 hours in May. This was even though the aircraft was removed from service for Project JUNIOR FLASH-UP from 23 March to 10 April 1960, and again in April for other modifications. Seven flights lasted over 10 hours and several hours were flown at supersonic speeds. Between 5-13 May the aircraft made seven sorties, five of them on successive days, proving that maintenance crews had significantly cut turn times. The tests proved the J79-GE-5A engines, with slightly improved performance over the -5 producing 9,700 pounds normal, 10,300 military, and 15,600 pounds thrust each with afterburners, performed well under real flight conditions.

Testing continued to prove that the B-58 was a very capable aircraft. B-58A (59-2428) underwent a very successful "cold soak" and quick reaction demonstration under Project WHITE HORSE during deployment to Ellsworth AFB, SD in January 1960. The demonstration was tagged with minimal maintenance in temperatures as low as two below zero and wind gusts up to 35 mph. The Hustler made three sorties during its stay at Ellsworth, each after an overnight cold soak. The crew was clocked during each flight to test take-off times under simulated combat conditions. The time needed to launch each flight after the cold soak was considered "very favorable" by the Carswell maintenance chief. Fuel, oil, and hydraulic leaks normally associated with cold weather operations were virtually non-existent. Support was provided in minimum time without mission delays. The entire series of demonstrations, which took a week instead of the one month planned, were conducted with minimum maintenance and ground support.

Category II testing was completed on 30 June 1960. Along with Category I testing, it achieved the Air Force objective of decreasing lead times and accelerating delivery of new weapon systems to operational units, having taken only about half the time of the previous eight-phase testing approach. It accumulated 1,216 flight hours and 256 sorties and progressed as planned. Notable during these tests were the first Mach 2.0 pod drops, and several endurance flights with the longest being over 18 hours long. Two aircraft performed stability and control evaluations at the AFFTC and CVCFW. Kirtland AFB, NM was used as a staging base for YRB-58A flights over the nearby Holloman AFB range where the first pod drops were conducted. The pods were designed by Convair to allow the B-58 to carry fuel and drop nuclear weapons. Testing was conducted by the Special Weapons Center (SWC) at Kirtland along with Convair and other experts. Convair was tasked with installing, arming, and fusing the warhead as well as flying the aircraft.

Convair YRB-58A takes off from CVCFW on 14 August 1958. (University of North Texas Libraries)

However, despite the many successes, seven test aircraft were lost between December 1958 and June 1960.

- On 16 December 1958, YRB-58A (58-1008) was lost near Cannon AFB, NM. The accident was attributed to a loss of control during normal flight when the auto trim and ratio changer became inoperative due to an electrical system failure.
- On 14 May 1959, YRB-58A (58-1012) was destroyed by fire after a fuel leak and accidental ignition during a ground refueling operation at CVCFW.

- On 16 September 1959, YRB-58A (58-1017) was destroyed when a tire blew during takeoff from Carswell AFB, TX.
- On 27 October 1959, YRB-58A (55-0669) was destroyed near Hattiesburg, MS when it lost control during normal flight with a Convair crew onboard.
- On 7 November 1959, YRB-58A (55-0664) was destroyed during a high-speed test flight near Lawton, OK when it disintegrated at high altitude. Convair Test Pilot Raymond Fitzgerald and Flight Engineer Donald A. Siedhof were both killed in this accident. They were attempting to collect side loads data of the vertical fin under conditions of the loss of an outboard engine at high speed.
- On 22 April 1960, YRB-58A (58-1023) crashed during flight test 29 miles NW of Ogden, UT after take-off from Fort Worth, TX with a Convair crew onboard. The crash was caused by loss of control due to air data system failure.
- On 4 June 1960, YRB-58A (55-0667) crashed near Lubbock, TX during flight test with a Convair crew onboard. The aircraft was enroute from Kirtland AFB, NM to Fort Worth, TX when the pilot lost control during normal flight. It was damaged by hail during a thunderstorm. All three crew members ejected but were killed when their open parachutes dragged through rugged terrain in high winds.

The accident rate made SAC apprehensive about the reliability of the aircraft in service and led to postponement of Category III testing. In addition, the 7 November 1959 accident raised questions about the complex flight control system. SAC insisted on correction of several deficiencies discovered through the Category II testing period including electrical malfunctions, tire failures, flight control difficulties, and possible structural weaknesses. The accident findings did not point to these as specific causes, but SAC had initiated flight restrictions pending further investigations and required certain modifications to improve safety. As a result, B-58s were restricted to subsonic flights only for nearly a year until the flight control system and tail structure could be modified.

The completed control system modifications led to more confidence in the aircraft and SAC speed restrictions were raised to Mach 1.5 in March 1960 for modified aircraft. Some structural modifications were also completed by mid-1960 (tail strengthened, critical side panels reinforced) and ARDC issued a report on 1 August 1960 indicating that there were no remaining design deficiencies in the aircraft or flight controls and that "when all functioned the systems met specs."[9] The report also indicated that SAC pilots verified good handling characteristics, but pilot training and high proficiency was necessary and flight control maintenance personnel had to be highly skilled. SAC accepted this report and assumed executive management of the B-58 program on the same day.

Convair began providing training at the CVCFW plant in March 1958 for Air Force maintenance personnel. Nine classes were completed in the first few months of 1959. Trainees were able to observe all equipment as it was tested, installed, calibrated, and operated, in addition to attending extensive classroom training. A Field Training

Detachment (FTD) assigned to Air Training Command (ATC) was established at Carswell in August 1960 to continue Air Force ground crew training.

Convair YRB-58A front view without pod. (National Museum of the U.S. Air Force)

SAC combat crews were prepared for the Category III Operational Test and Evaluation (OT&E) role by the 3958th OETS during Category II testing. The 3958th began formal three-month classes in May 1960 with no personnel difficulties and the 43rd Bombardment Wing (BMW) "saw no fault" in 3958th performance. The B-58 was considered the most technically advanced bomber of its day and required the most highly qualified crew members to fly it. The selected students were for the most part former B-47 crew members, highly qualified, and recommended by their wing commander with a minimum of 1,000 hours flying experience in SAC multi-engine jet bomber-reconnaissance aircraft and 500 hours since upgrading to aircraft commander. But even this did not guarantee acceptance as an elite B-58 crew member. Each student was expected to complete a grueling training program, and some were washed out before completing the requirements. Initially, they used the Convair two-seat TF-102A for training until the TB-58A became available in August 1960. Students typically flew three TB-58A flights before being qualified to solo in the B-58A. Each TB-58A typically flew three missions during every 24 hours.

Category III OT&E testing was conducted by the 43rd BMW and started on 1 August 1960 at Carswell AFB.[10] Category III testing was primarily focused on the development of techniques and procedures for using the B-58 most effectively as a combat weapon. It had three primary objectives: testing and evaluating tactical employment of the integrated weapon system, training aircrews, and maintaining combat readiness of the 43rd BMW. It evaluated overall B-58A operational performance and tested all aircraft systems including passive defense, electronics, communications, and navigation as well as Aerospace Ground Equipment (AGE) and supply availability (all played a part in operational success). The highly integrated systems on the B-58A meant broad testing but there was critical focus on the AN/ASQ-42 BNS.

SAC assumption of B-58 executive management and the start of Category III testing marked several changes. The ARDC 6592nd TS was inactivated with its assigned YRB-58A aircraft and personnel transferred to the 43rd BMW. The B-58 JTF was officially dissolved, but a small group of ARDC personnel stayed at Carswell to assist the 43rd BMW during Category III tests. These changes gave SAC and the 43rd BMW the triple mission of testing the aircraft, training crews, and assuming a SAC combat mission. This had an unfortunate psychological effect on some Air Force and SAC officers who were accustomed to receiving production weapon systems that were thoroughly wrung out (even though some production modifications were still ongoing). SAC's early experience with the B-58 was full of frustration and disappointment. Many people mistakenly expected the aircraft in its operational testing to be as trouble-free as the B-47 or B-52. It was not.[11]

But the aircraft continued to perform well and achieved some significant milestones. For example, a B-58A (59-2431) from Carswell flew at Mach 2.0 for 78 minutes in October 1960. This was believed to be longest Mach 2.0 flight at that time.[12] The aircraft flew north at Mach 2.0 for 32 minutes, refueled, and then flew southward for 46 more minutes. The total flight lasted three hours and the aircraft flew at supersonic speeds for 1 hour and 48 minutes. Most of the supersonic flight was above 50,000 feet. The aircraft was piloted by Lieutenant Colonel J.F. Cotton, ARDC special assistant to the 43rd BMW commander. The crew was also comprised of Major Andrew Rose, Jr. (navigator/bombardier) and Captain Samuel J. Doren (DSO). The flight was "smoother and quieter" than usual according to Rose. The crew added to their stellar performance record on October 17 when they flew Mach 2.0 for 31 minutes during daylight and 26 minutes at night, marking the longest nighttime Mach 2.0 flight at that time.

Also, a 43rd BMW B-58A completed a 12 hour and 45-minute sortie over polar latitudes. The flight was undertaken to evaluate the operation of the search and doppler radar systems and the BNS operation at high geographic latitudes. The crew consisted of

Colonel J.K. Johnson, 43rd BMW commander (pilot), Captain W.L. Polhemus (navigator/bombardier), and Captain R.R. Wagener (DSO).

The Category III OT&E program ended on 31 July 1961 when B-58A (59-2435) flew the last flight test. A total of 5,265 flying hours were logged during the OT&E program with 945 strictly for test, 1,879 training hours, and the remaining 2,439 were for maintenance test flights, acceptance and delivery flights (new and retrofitted B-58A aircraft), airshows, record breaking flights, and ferry missions.

Convair B-58A (59-2435) taxis out for the last B-58 flight test. (University of North Texas Libraries)

SAC initially used the YRB-58A aircraft transferred from ARDC for Category III OT&E. As the flight test program neared completion, the Air Force was faced with the problem of what to do with these flight test aircraft. Many of them had low times on their airframes and were hence still viable from a useful life standpoint. On 20 June 1958, the B-58 project office made a presentation to the Air Force which stated that the first 13 aircraft were suitable only for test work. These aircraft were limited by structure to a 147,000-

pound gross weight and were not convertible to a tactical configuration without major modification. The final 17 aircraft, however, were built to a 158,000-pound gross weight requirement and, if the landing gear were replaced, could have a 163,000-pound (normal) capability.

Consequently, the Air Force decided to retrofit these aircraft to the B-58A production standard for operational service under project JUNIOR FLASH-UP. The first phase of this project started on 15 February 1960 and affected five aircraft from the Carswell AFB test force (55-0671, 58-1013, 58-1014, 58-1019, and 58-1020). Each aircraft received a 10-day "facelifting treatment" aimed at providing the Air Force with current configuration changes and reducing maintenance costs. YRB-58A (58-1013) was the first aircraft completed two and a half days ahead of schedule. YRB-58A (55-0671) was the last aircraft completed on 9 April 1960, almost two weeks ahead of the planned 22 April completion date.

Eventually, eleven of the 17 YRB-58A (58-1009 to 58-1011, 58-1013 to 58-1016, 58-1018 to 58-1021) aircraft produced under the second contract were upgraded in Phase II of project JUNIOR FLASH-UP.[13] This phase was much more involved than Phase I. The changes needed for each aircraft were different since they were instrumented for different phases of the test program. Major upgrades included installation of improved J79-GE-5B engines and the automatic center of gravity system. All systems for each aircraft were upgraded to the final production configuration. The upgraded aircraft were redesignated as B-58A and were fully interchangeable with production B-58A models. The first converted B-58A (58-1021) was delivered to the 43rd BMW at Carswell AFB, TX on 30 December 1960. The last converted B-58A (58-1015) was delivered to the 43rd BMW on 25 October 1962.

Operational Aircraft

B-58A. The B-58A would ultimately see service with SAC despite the command's resistance to it. The first operational B-58A (59-2428) was ready in the spring of 1959. The first flight occurred in September 1959, and the aircraft was delivered to the 6592nd TS at Carswell AFB, TX for R&D testing on 1 December 1959. It was also the first B-58A delivered to the 43rd BMW on 1 August 1960 making it the Air Force's first operational Mach 2 bomber although some tactical systems were not installed. The first fully operational B-58A (59-2436) equipped with all tactical systems was delivered to the 43rd BMW on 3 September 1960. The first B-58A (59-2461) assigned to the 305th BMW at Bunker Hill AFB, IN was delivered on 11 May 1961. Production ended in the fall of 1962 and the last three deliveries of B-58A (61-2078, 61-2079, and 61-2080) aircraft were completed at the 305th BMW on 26 October 1962, which was one month ahead of schedule. This was also the same day that the last B-52H (61-0040) was delivered to the 4136th Strategic Wing (SW) at Minot AFB, ND. This marked the end of SAC bomber production for the first time since 1946. There were also no other bombers in development since the XB-70 was excluded from consideration as a production bomber one month earlier in September 1962.

B-58A Assembly line at CVCFW on 3 March 1960. (University of North Texas Libraries)

The B-58A became operational in 1961 but did not achieve the planned numbers or capability.[14] Only 86 of the 116 aircraft purchased were considered production, while the first 30 were R&D prototypes. In addition, the activation of the first operational squadron was delayed from June to December of 1960 and the first wing of 36 rather than 45 aircraft would not be ready until August 1961. The lack of funds, competition from other weapon systems, and a variety of complex and at times biased, political and technological decisions had all combined to cause delays in B-58 operational deployment.[15]

Due to the delays in the B-58A production program, the various aircraft that had been delivered had great variations in equipment, systems updates, maintenance requirements and capabilities. As a result, the Air Force instituted project SENIOR FLASH-UP for the first 36 production aircraft (59-2428 to 59-2463) to update and retrofit the aircraft to the current production configuration.[16] Among the changes introduced were anti-icing systems, rendezvous beacons, position indicating beacons, long range communications, emergency UHF, provisions for escape capsules, ECM chaff dispensers, improved HACON[17] long range communication system, TACAN installations, modernizing the BNS, refurbishing the crew compartments, modernizing and standardizing the flight control system, and a structurally improved vertical fin and fuselage empennage systems.[18]

Six Hustlers undergoing project SENIOR FLASH-UP modifications in the Experimental Building in May 1962. (General Dynamics News, 9 May 1962)

The modifications were completed in two "cycles" with Cycle I running from September 1960 to May 1961. The first eight aircraft were cycled through the program for 30-day periods and the remaining aircraft for 12 days each. Two aircraft were in work at the same time throughout the program. B-58A (59-2434) was the first aircraft to enter SENIOR FLASH-UP Cycle I on 2 September 1960. Cycle II ran from April 1961 to August 1962. The number of modifications required varied by tail number and some aircraft were included in both cycles. The aircraft averaged 6-12 weeks in the program depending on the modifications required. B-58A (59-2429) was the first aircraft to enter the SENIOR FLASH-UP Cycle II modification program on 21 April 1961. B-58A (59-2463) was the last aircraft to complete Cycle II on 9 August 1962.

The B-58A was unique among SAC bombers in that the crew consisted of only one pilot, a navigator/bombardier, and DSO. The crew members all sat in individual compartments seated in tandem vs. the side-by-side seating preferred by General LeMay and other SAC leaders. Structurally, the three crew compartments comprised a single pressurized cabin, but structural bulkheads and equipment between them created a compartmentalized effect. Each compartment had a separate canopy hinged at the rear for entry and exit. The compartments prevented direct vision or physical contact between crew members during flight. A crawlway between the pilot's station and the second crew station on the right side of the fuselage could only be used for maintenance of electronic equipment, but a crawlway between the second and third crew stations could be used for passage during flight. However, the crew was essentially strapped in their seats as they would be in a fighter making the crawlways useless; a "clothesline" system was fitted along the cabin wall to allow them to pass notes or small items back and forth.

Convair B-58A (59-2451) highlighting the three tandem crew compartments. (U.S. Air Force)

The pilot had a windshield with six adjacent panels, plus one panel on each side of the canopy. This afforded excellent outside vision, and the pilot could see parts of the exterior of the aircraft as well as the engine nacelle inlets.[19] All gauges, switches, knobs, and controls were within easy reach of the pilot. Four throttles, one for each engine, were in a quadrant on the left side of the pilot's station. Flight instruments were on the forward left panel and engine instruments on the right forward panel. The lower left panel had the gages to monitor the two hydraulic systems.[20] The right side contained fuel control panels, where fuel could be shifted to set the aircraft Center of Gravity (CG) depending on the mode of flight, subsonic or supersonic. Setting a CG would automatically shift fuel to the desired setting.[21]

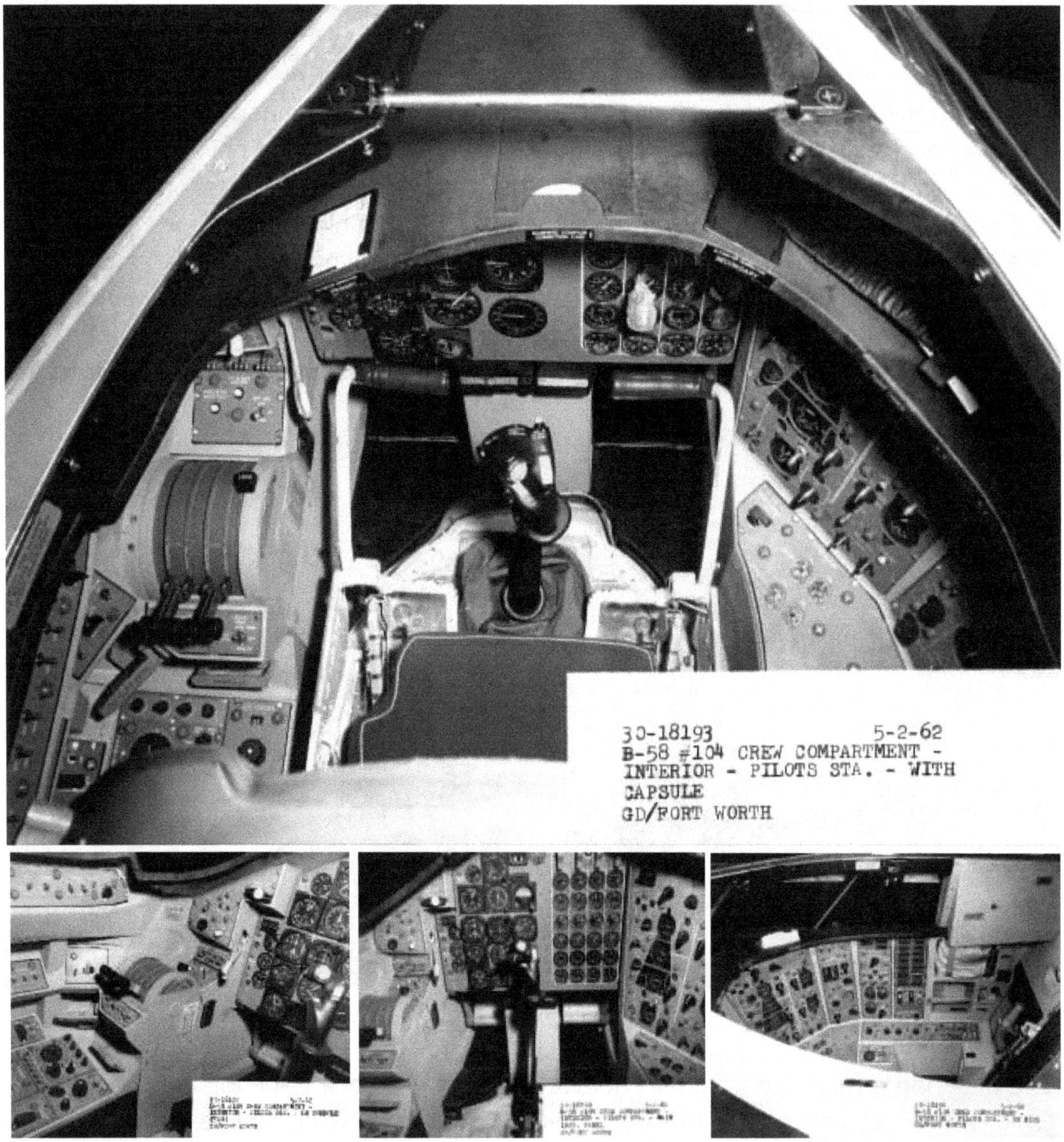

Convair B-58A pilot station with crew escape capsule installed. (National Museum of the U.S. Air Force)

Behind the pilot, the navigation computer was housed within the pressurized cabin to help keep it within operational temperature limits during supersonic flight. Next came the navigator/bombardier and the DSO compartments; each only had small side windows, one on each side. The navigator/bombardier's panel was equipped with bomb and pod dropping instrumentation, bombing system indicators and monitors, plus the navigation equipment. The DSO's panel contained passive and active defense system monitors.[22]

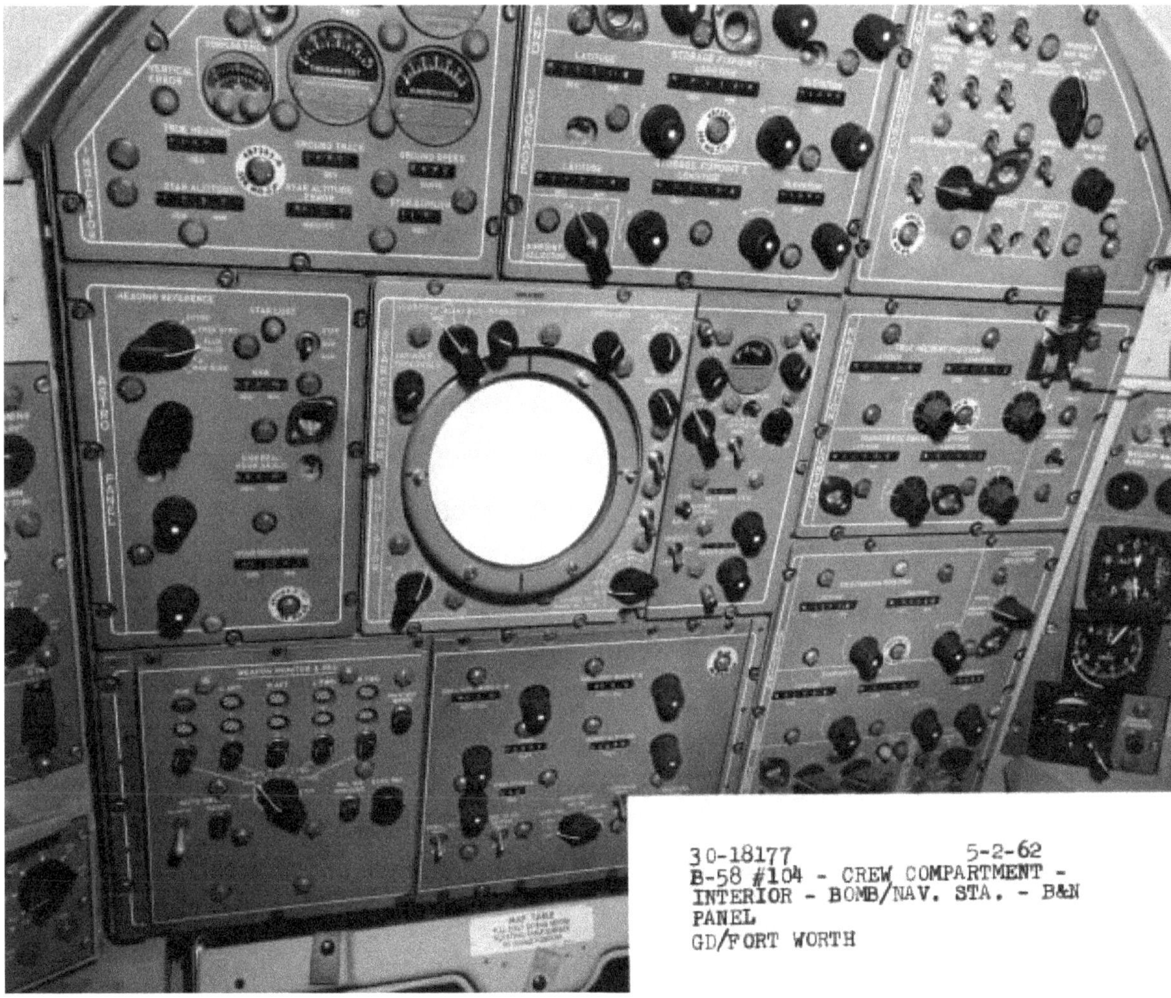

Convair B-58A navigator/bombardier station BNS panel. (National Museum of the U.S. Air Force)

The delta wing configuration required a unique control surface configuration. There were no horizontal stabilizers, flaps, ailerons, or elevators like on typical bomber aircraft. Instead "elevons" installed on the trailing edge of the delta wings controlled the aircraft. The pilot controlled the aircraft using a stick, like a fighter aircraft, as opposed to the control wheel used in other SAC bombers. The flight control system provided the pilot with a constant feel on the stick whether flying subsonic or twice the speed of sound.

Forward and aft stick movement caused the elevons to move together for pitch control, while left and right movements caused them to move in opposite directions for roll control. Each wing control surface and the rudder were controlled by separate hydraulic systems. A triple redundant yaw damper system stabilized the rudder and provided opposite rudder deflection in the event of an engine loss at supersonic speeds.

The flying qualities of the aircraft were improved by various features including the elevator ratio changer, automatic trim, aileron ratio changer, aileron-rudder interconnect, wing heavy control, and roll, pitch, and yaw stability augmentation. The ratio changers, elevator and aileron autopilot and dampers servos, elevon mixer, trim motors, artificial feel systems, and the aileron-rudder interconnect were grouped together in the Power Control Linkage Assembly (PCLA). All pilot-initiated commands were routed through the PCLA to the control surface hydraulic valves. All other commands, except the rudder damper servos commands, were initiated by the PCLA components.

B-58 in-flight bottom view with MB-1C pod highlighting the delta wing and control surfaces. (University of North Texas Libraries)

The elevon and rudder flight control surfaces were controlled by the pilot, autopilot, and stability augmentation system (dampers) and utilized hydraulic actuation. The elevons performed the combined functions of the ailerons and elevators and were controlled by conventional stick commands. An elevon mixer assembly mechanically converted the elevator and aileron commands into right and left elevon commands. The rudder was controlled by conventional rudder pedals and the yaw damper, plus the action of the aileron-rudder interconnect.

The autopilot and stability augmentation system (dampers) computers, sometimes called the "electronic brain" of the flight control system, continuously monitored Mach number, altitude, attitude, acceleration, gross weight, and various other conditions. They then processed this information and passed data to the PCLA components to initiate control surface movements. The system was designed to assist the pilot throughout flight and achieve optimum stability while permitting the necessary maneuverability. When placed in the automatic flight mode it also limited control surface movement at higher airspeeds and precluded the aircraft from exceeding the maximum G-force limits, thereby preventing the pilot from making any violent maneuvers that could overstress the aircraft.

Convair B-58A assembly line at CVCFW. Notice the aircraft internal structure highlighted in the foreground. (University of North Texas Libraries)

The airframe was unusual and consisted of hundreds of internal traverse corrugated duralumin spars running spanwise through the fuselage at intervals of 11-15 inches. There were no chordwise ribs except for control surface, undercarriage components, and engine pylon attaching points. The B-58 was the first to use a bonded aluminum honeycomb structure as the inner and outer panels for the entire wing and tail surfaces and most of the fuselage. This gave it a light but very strong and stiff structure with a smooth exterior surface, further enhancing its stealth properties. The wing skin panels used a Convair developed bonded thermal sandwich featuring "top and bottom layers of duralumin alloy sheets with a half-inch thick honeycomb core of either phenolic-resin-fiberglass cloth or light gauge duralumin" attached with titanium screws. The fuselage panels were reinforced with beaded inner skins bonded to outer skins. Hot spots, including the engine exhaust panels and elevons, used brazed stainless-steel sandwich panels.

Much of fuselage structure, entire wing, and external weapons pod (both types) contained integral fuel tanks (approximately 100,000 pounds total). Four tanks were contained in the wings, with the largest 5,893-gallon capacity (38,306 pounds) in the aft portion of the wings (1) and 3,202-gallon tanks (20,811 pounds) in the forward portion of the wings (2). The fuselage tanks were smaller and included a 610-gallon reservoir tank above the wing (3) and 1,219-gallon balance tank (4) in the rear (total 11,888 pounds). The reservoir tank was an accumulator tank which used an auto transfer

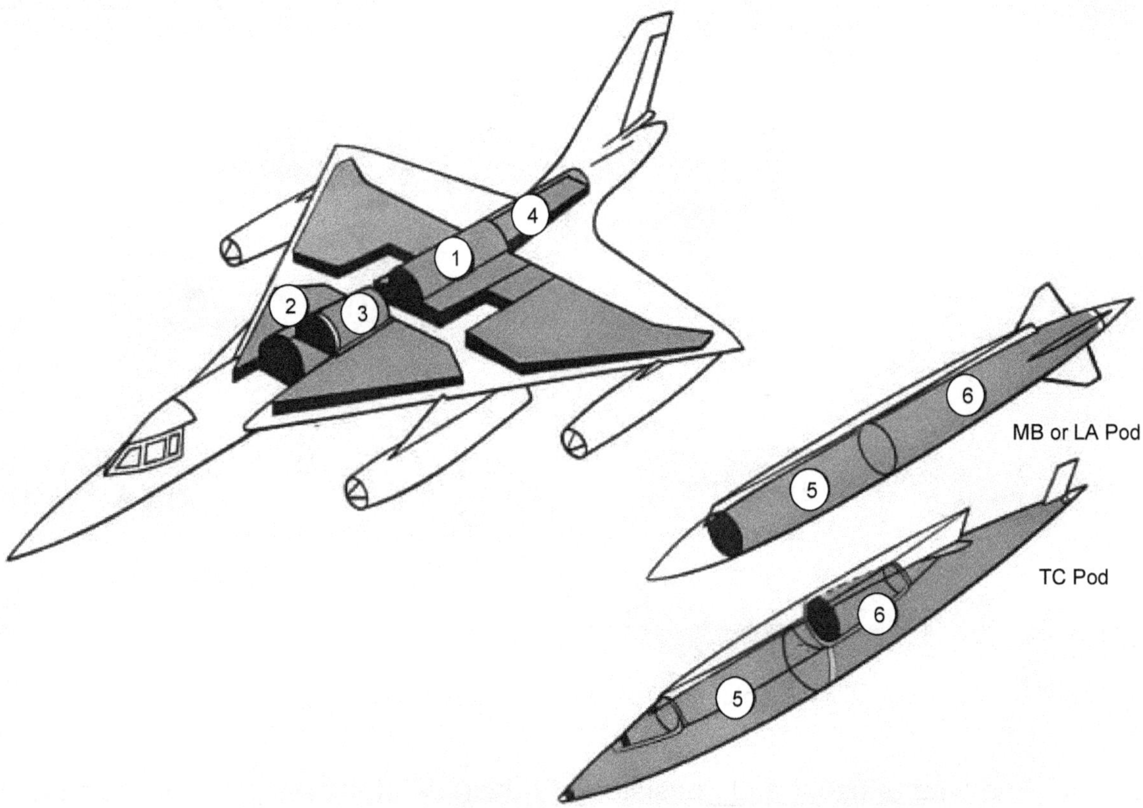

Much of the B-58 fuselage structure, entire wing, and external weapons pod (both types) contained integral fuel tanks capable of holding approximately 100,000 total pounds of fuel. (U.S. Air Force)

system to maintain a specified tank level until the other tank supplies were depleted. The balance tank did not provide direct engine supply, but fuel could be transferred from the balance tank to the forward tank when needed. Fuel could also be carried in the under-fuselage pods (5, 6) with 4,172 gallons (27,121 pounds) in the MB-1C or LA-331 pod and 3,885 gallons (25,254 pounds) in the TC pod. The fuel system could operate with or without the pod being installed.

The fuel system was the most complex and sophisticated installed in an operational aircraft due to the tight space restrictions, the problem of fuel "stacking", and complex CG requirements to keep the airplane properly balanced. Automatic in-flight pumping of the fuel maintained the aircraft CG for various airspeeds. Manual fuel management responsibility was shared between the pilot and DSO. The aircraft could be refueled in-flight to a gross weight of 176,890 pounds. A "high-speed" modification called the HEAVYWEIGHT RETROFIT PROGRAM performed from late 1961 to early 1962 on 34 B-58A and two TB-58A aircraft allowed them to take-off with up to 176,890 pounds gross weight including fuel. Aircraft without the modification were still limited to 163,000 pounds gross take-off weight.

The B-58A possessed many revolutionary features including innovative construction methods and materials. Several components were fabricated using a new extremely light and heat-resistant alloy called Magnesium-Thorium (Mag-Thor). This alloy was developed from magnesium and thorium extracted from the seas and sea-washed sand. When combined, the two metals produced an alloy that was one third lighter, more rigid, and much more heat resistant than aluminum. However, Convair metals experts had to perform many experiments, with many failures, to develop processes for forming the metal without it breaking and warping. Eventually, they succeeded, and several Mag-Thor parts were used for critical high temperature applications along with the other "wonder metal", Titanium.

The four installed J79-GE-5A or -5B engines produced a combined thrust of 62,400 pounds with afterburners. The J79 was an axial-flow turbojet with a 17-stage compressor and a three-stage turbine. It was equipped with variable-position inlet guide and stator vanes in the first six compressor stages and a variable exhaust nozzle. The inlet guide and stator vanes were automatically programmed for speed and temperature to minimize compressor stalls by admitting the correct volume of air and channeling it to the compressor vanes at the required Angle of Attack (AoA). The inlet guide and stator vanes were mechanically linked together externally to operate in unison. They were automatically modulated by the engine fuel control unit as a function of engine speed and compressor inlet temperature. The inlet guide and stator vane control unit, an integral part of the engine fuel control unit, directed high pressure fuel to actuators which hydraulically positioned the vanes to their scheduled position. A mechanical feedback linkage eliminated vane angle error and overtravel of the vanes.

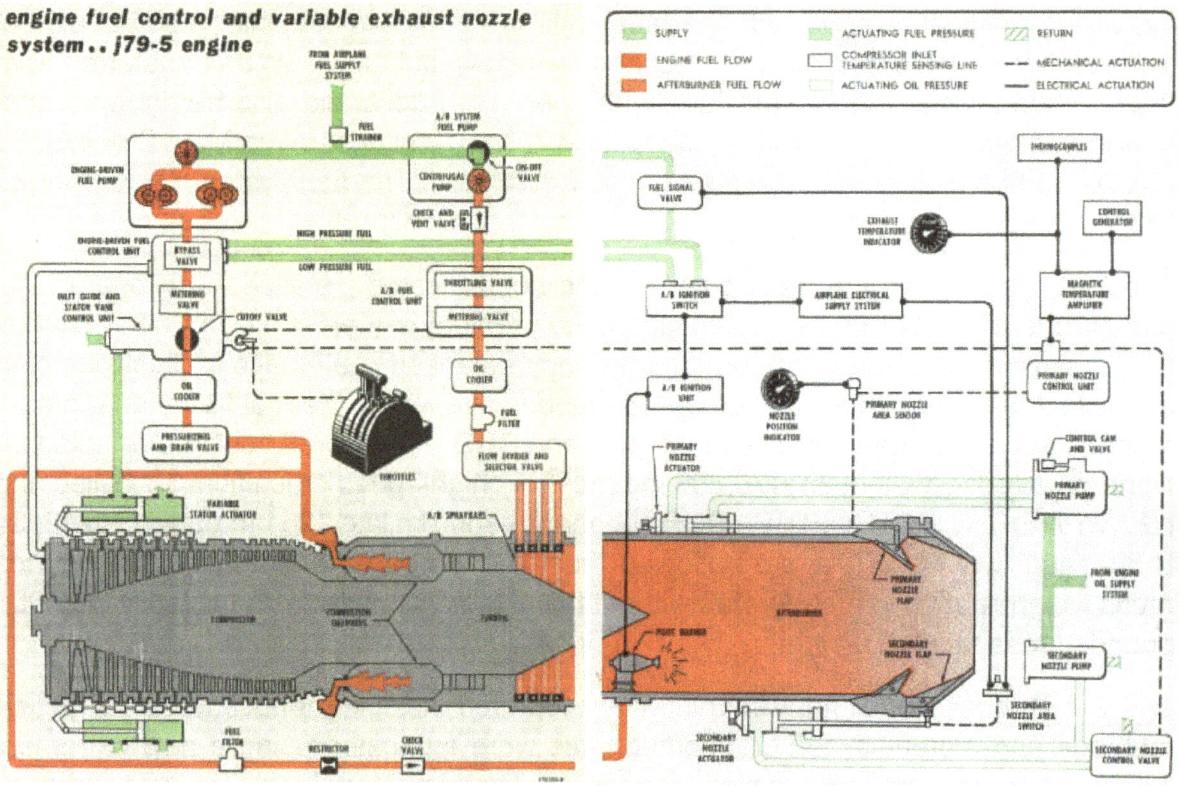

The J79-GE-5 engine was equipped with variable-position inlet guide and stator vanes in the first six compressor stages and a variable exhaust nozzle. The inlet guide and stator vanes were automatically programmed for speed and temperature to minimize compressor stalls. (U.S. Air Force)

The variable exhaust nozzle system automatically controlled the exhaust area to provide for optimum thrust and specific fuel consumption under varying engine operating conditions. It also protected the engine from overheating. The system consisted of primary and secondary nozzle flaps, plus associated control units and actuators. Each primary nozzle control was mechanically interconnected with the engine throttle and synchronized so that throttle movement would automatically result in the proper primary engine nozzle setting. The secondary nozzles were used to provide maximum thrust and reduced drag during cruise and military operating ranges. These were opened during idle and afterburner operations and were closed for operations in the cruise and military ranges.

A variable position inlet spike was located in each engine nacelle to maintain an efficient inlet airflow throughout the speed range of the aircraft. The B-58 was one of the first aircraft to implement this technology. The purpose was to maintain a subsonic airflow to the engine when the aircraft was flying at supersonic speeds. The inlet spike operated by sensing and comparing the outside air pressure at the tip of the spike to the air pressure on the inner surface of the inlet lip and then moving the cone to maintain a constant ratio between the two readings. The spike remained in the aft or retracted position until Mach 1.42 was reached, at which time the ratio was maintained by moving the spike progressively forward as Mach number increased and backward as Mach number decreased. Two things occurred. Along with restricting the airflow, the spike prevented shock waves from entering the engine air chamber by forcing the shock wave to a position just forward and above the engine inlet cowling. The system was highly reliable and fully automatic, not even needing to know the speed at which the B-58 was flying. It only needed to know the ratio of outside air pressure to inside air pressure and then adjusted the spike as necessary to maintain the optimum ratio.[23]

Convair B-58A (60-1118) in flight. Note the variable position inlet spikes in the engine nacelles. (U.S. Air Force)

The landing gear raised the aircraft about 10 feet off the ground to make way for the underbelly pods and was designed to handle 165 knots touchdown speed at 75,000 pounds gross weight. The landing gear retracted into pressurized and air-conditioned wheel wells to prevent the tires from melting due to the high heat generated at Mach 2.0 speeds. Each main truck had four split-type "non-frangible" steel wheels, and each wheel had two 22-inch tires inflated with nitrogen at 240 pounds per square inch (psi).

B-58 main landing gear with four split-type "non-frangible" steel wheels, and two 22-inch tires on each wheel. (University of North Texas Libraries)

The aircraft could roll safely on the steel wheels in the event of tire failure. The non-frangible wheels were installed beginning in 1961 after some B-58 aircraft experienced multiple tire failures during take-off and landings. The wheels incorporated a one-inch-wide center rolling flange of extremely strong solid aluminum located between two 22-inch tires mounted on magnesium-aluminum flanges. The center flange prevented the inner and outer wheel flanges from contacting the ground if both tires became underinflated. This center flange resisted wheel break-up in case of a multiple tire failure on take-off. It also incorporated two fusible plugs installed in the inboard wheel half to relieve tire inflation pressure and lessen the possibility of tire blow-out caused by excessive brake energy absorption during landing. The nose gear had two tires which were fully steerable. The nose gear had a unique retracting mechanism that slid aft while folding forward allowing it to clear the weapons pod.

Alternating Current (AC) power was provided by three 40 kilovolt-amp (kVA), 111 ampere, engine mounted generators (115-200 VAC) driven by engines No. 1, 2, and 3. Each generator had a constant speed drive which maintained the generator speed at 8,000 revolutions per minute (rpm). Decouplers were provided to decouple a malfunctioning constant speed drive from its generator. In normal operation, the No. 1 generator powered the left main AC bus, and the No. 3 generator powered the right main AC bus. The No. 2 generator served as a standby power source on the bus crossfeed circuit.

Direct Current (DC) power was distributed by a network of DC buses which received power from eight multiple-voltage DC transformer-rectifier power units and a 28-volt battery for emergencies. The transformer-rectifier units converted AC power into DC power. The eight units were equally divided into a forward and aft bank, and the individual units in each bank were paired together and wired in parallel. The paired units could produce 250-volts DC, 150-volts DC, negative 150-volts DC, or 28-volts DC. The 28-volts DC units could also produce 28-volts AC power for miscellaneous low-voltage AC requirements. Each unit was marked with the voltage output. If a unit in either bank failed, the corresponding unit in the other bank assumed the full load for that voltage.

Two independent hydraulic systems (primary and utility) with two engine-driven pumps generated 3,000 psi. Hydraulic power for the elevon and rudder control surfaces was equally divided between the primary and utility systems, with each system connected to separate actuators. Each system could assume the entire control system load if one failed, although the maximum control surface force output was limited to half of the normal dual system operation. Total loss of both systems rendered the aircraft uncontrollable. The utility system also supplied hydraulic power for the landing gear, nose wheel steering, wheel brakes, tail turret, search radar, chaff dispensing system, air refueling system, aileron damper servo, elevator damper servo, autopilot servos, and rudder damper servo. The only additional equipment receiving hydraulic power from the primary system was the rudder damper servo.

Life support included two separate air conditioning plants (crew and electronics cooling). The air conditioning system used bleed air from the 17th stage compressor discharge of the No. 2 and 3 engines to provide conditioned air for crew compartment and electronic equipment cooling, heating, pressurization, and ventilation as well as windshield rain removal and defog, engine and inlet spike anti-icing, and fuel tank pressurization. System operation was fully automatic throughout normal flight. The air conditioning plants provided both cool and warm air, allowing the pilot to select the desired cabin temperature and at the same time prevent overheating the electronic equipment. The same air automatically furnished cabin pressurization as selected by the pilot.

The Magnavox AN/ARC-57 UHF Command Radio and communication system provided a means of crew intercommunication, plus normal and emergency UHF air-to-air and air-to-ground communication. The AN/AIC-17 Ground Interphone was connected to the external interphone receptacle to enable ground crew communications. The AN/ARC-34 UHF Command Radio provided back-up air-to-air and air-to-ground communication. The AN/ARC-74 UHF Emergency Communication System provided back-up communication capability in the event of a command radio failure. The AN/ARC-68 or AN/ARC-110 provided standard long-range communications. The Huges Aviation HACON AN/ARC-87 Long Range Communication system provided air-to-air and air-to-ground HF communication at a range of up to 5,000 miles. The long nose boom included a pitot tub e on the end and also operated as an HF probe antenna.

Radio navigation systems included an AN/APX-47 air-to-ground IFF system, and AN/APX-48 air-to-air IFF system, ARN-50 VHF navigation system, AN/ARN-65 TACAN,

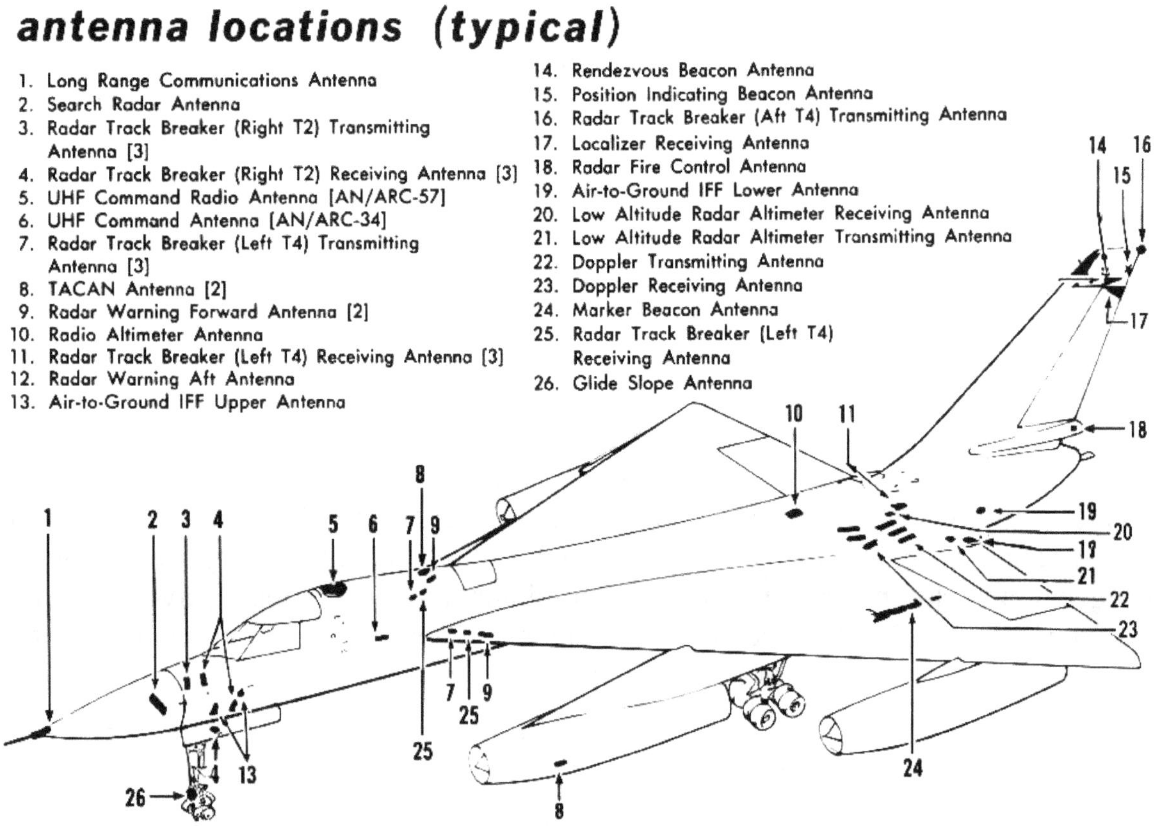

The B-58 aircraft had an array of communications and navigation equipment requiring numerous antennas across the aircraft. (U.S. Air Force)

and AN/ARN-69 Instrument Landing System (ILS), as well as AN/APN-136 Position Indicating and AN/APN-135 Rendezvous Beacons. An AN/ASH-17 punch-tape inflight recorder was installed for the first time in any aircraft to record flight information for inflight reference and post flight analysis. An AN/ASH-15 Bomb Damage Evaluation System consisting of a camera and data package was also installed. In conjunction with the inflight recorder and the Primary Navigation System (PNS), it recorded information to derive weapon ground zero. The aircraft included a sophisticated new audio warning system (using female voice), fitted midway through the aircraft's operational service, and a visual warning and caution lamp panel.

The autopilot was very sophisticated and included a "watchdog" function that prevented the pilot from making maneuvers that would be dangerous in supersonic flight. Pilot selectable modes of operation were attitude stabilization (basic autopilot), constant Mach, constant Mach-altitude, navigation ground track, and constant ground track. The pilot could control the aircraft without disengaging the autopilot through the control stick steering function and the navigator/bombardier could control the aircraft during bomb runs. The autopilot received inputs from the Air Data Computer as well as the PNS and the Auxiliary Flight Reference System (AFRS). The AFRS consisted of several components including the MD-1 Vertical Gyro, Rate Switching Gyro, and the J-4 Magnetic Compass system which included a Directional Gyro, Servo Amplifier, and Remote Compass Transmitter (Flux Valve).

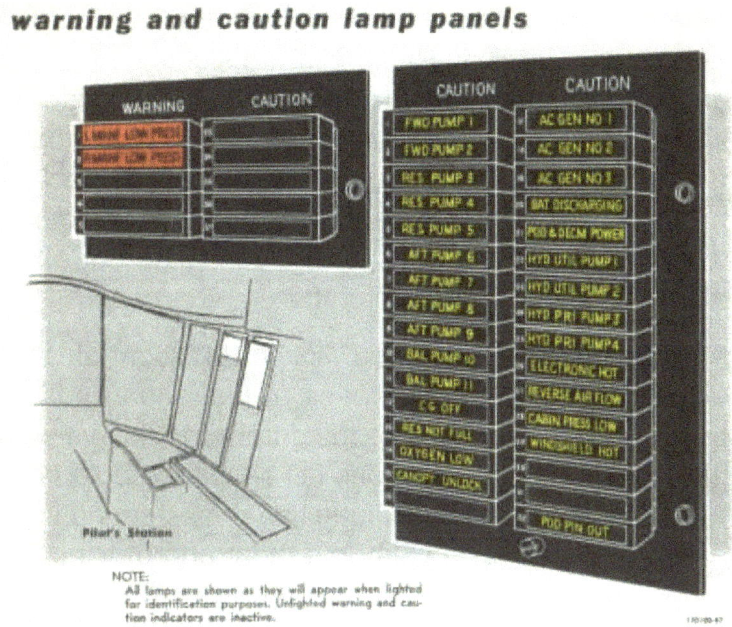

The B-58 included a sophisticated new audio warning system (using female voice) and a visual warning and caution light panel. (U.S. Air Force)

Despite the several advancements contained in the B-58A it experienced several initial and recurring problems. The first 47 aircraft were fielded without a TACAN installed. The units were provided as GFE and late delivery required retrofit of the aircraft. The wheel and tires had previously caused the loss of a B-58A on 16 September 1959 (destroyed by fire after aborted take-off at Carswell) and were being upgraded with new sturdier units. Perhaps most significant, even though the B-58 could fly low-level missions it lacked equipment that would allow it to fly even lower such as a highly accurate Radar Altimeter and Terrain Following Radar (TFR). Design changes were planned and authorized in mid-1959 but required time to implement. Unfortunately, they were never fully carried out due to funding issues.[24]

The AN/ASQ-42 BNS, also called the "Stellar-Inertial" navigation system, was extremely complex and its initial poor reliability was a problem for the B-58. The B-58 was the first aircraft to feature such a system. The system used 1950s vacuum tube technology. It weighed 1,200 pounds and was comprised of 19 interconnected and highly integrated black boxes, which made it very difficult to troubleshoot. The system was developed by Sperry Gyroscope Company and was based on systems it had designed for use as missile guidance systems.

The BNS included the PNS comprised of a highly accurate Inertial Measurement Unit (IMU) known as the "Stable Table" consisting of gyros and accelerometers to sense attitude and acceleration. An AN/APN-113 Doppler Radar system sensed speed and drift to provide ground speed and wind data. It provided a considerable advancement and simplification in doppler radar using Ku band continuous wave (rather than pulse) technology and fixed (rather than space stabilized) antennas.[25] A KS-39 Astrotracker system provided an accurate heading reference to the Stable Table by tracking the Sun,

Moon, or stars. Stars could even be tracked in daylight.[26] A high-resolution Ku Band Search and Tracking Radar allowed sighting of known fixpoints for present position updates and direct target sighting for bomb release. It could also provide a continuous view of the terrain ahead and capture radar photographs. It was the highest-resolution bombing radar then in tactical use with less space, weight, and frontal area. Both the pilot and navigator/bombardier had a Radio Altimeter to provide altitude above terrain with an accuracy of +/- 250 feet from 500 feet above the terrain up to the maximum altitude capability of the aircraft.

An Integrated Computing, Display, and Control Unit tied together all sensing units. It also provided alternate modes of operation and sensors. For example, if the Astrotracker failed, magnetic heading from the flux valve could be switched in; should Doppler Radar speed indications fail airspeed could be used. A tracking handle was used for radar longitude and latitude updates, steering the aircraft on bomb runs (through the autopilot), and making airborne radar approaches.

The BNS also included the Bombing System which was used in conjunction with the PNS to steer the aircraft on the bomb run and to automatically drop weapons at a time calculated to hit a predetermined target. The weapons controlled consisted of a pod, and four small weapons (Mk-43 or B-61 bombs) mounted in tandem, two on each side of the pod. The Bombing System consisted of a ballistics computer and necessary controls. The ballistics computer was an analog computer which received signals from the PNS, Air Data Computer, and data set into the control panel on ballistic winds and desired burst altitude. The ballistics computer computed and supplied trail and time of fall signals to the PNS. The PNS used these signals received from the ballistics computer to compute steering error signals and an automatic release signal.

The AN/ASQ-42 BNS was considered SAC's most sophisticated BNS and was designed to find and bomb any target from high-altitude/supersonic or low-altitude/subsonic flight profiles. It was estimated to be ten times more accurate than previous systems. In Navigation mode, the desired coordinates that the navigator/bombardier entered into computer and flight instruction were sent to the autopilot. The nose-mounted search radar and display units constantly updated position, terrain, and altitude data and indicated when pre-set locations were reached. In Attack mode, the autopilot would fly the aircraft to the target along pre-set fix points (displayed on the radar scope). The navigator/bombardier could make small course corrections and weapons release was automatic (to ensure airburst over the target). In either mode, the search radar could be set to non-radiating standby mode to reduce aircraft radar cross-section during low-level penetration missions.

The system provided the crews with aircraft attitude and heading; present position; continuous ground speed reading; semi-automatic correction of present position; semi-automatic calibration of altitude and airspeed; automatic tracking of stars to assure accurate headings; accurate polar navigation; automatic radar bombsight crosshair laying; advance storage of fix points and offset points; automatic radar photography; a map comparer to aid in target identification; automatic ballistic computation for weapons drops; and in-flight malfunction detection and correction.

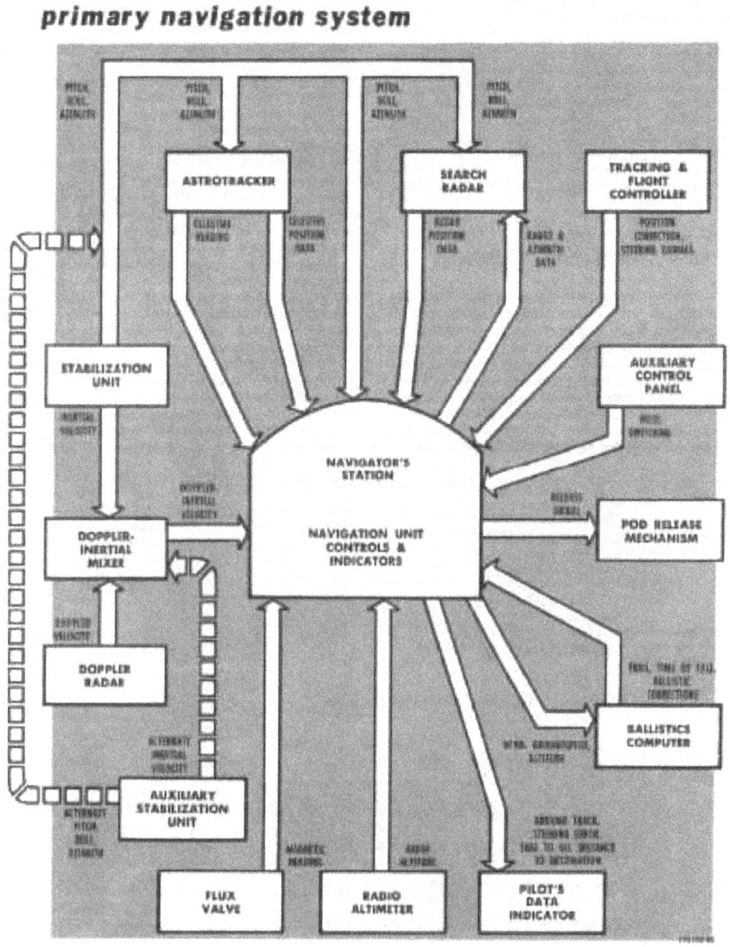

The AN/ASQ-42 BNS was considered SAC's most sophisticated BNS and estimated to be ten times more accurate than previous systems. (U.S. Air Force)

The ejection seats used in YRB-58A aircraft were initially retained in the B-58A and remained unsafe. These seats, designed and produced by Convair and dubbed the "SACseat", used a ballistic-initiated, rocket-catapult ejection system which operated independently of other aircraft systems. The backrest of the seat provided storage for a backpack type parachute, and the bucket portion of the seat provided storage for the survival kit. A gas operated human-seat separation system facilitated separating the crewmember from the seat once clear of the aircraft.

However, the seats failed to produce enough thrust to clear the aircraft during supersonic flight, which resulted in the death of one crewmember and injury to the remaining two during flight testing on 15 December 1958. The seats were modified as an interim measure in 1959 to add restraints to keep the crew members arms and legs tightly in place during ejection and protected from high-speed air blasts capable of breaking bones. The crew member stayed with the seat until it cleared the aircraft and just before the parachute opened, at which time the restraints were automatically relaxed, and the seat was jettisoned.

Initial testing of the SACseat was accomplished using the "supersonic sled run" at the Supersonic Naval Ordnance Research Track (SNORT) in China Lake, CA.[27] The sled

Sled used for B-58 SACseat ejection seat testing. (National Museum of the U.S. Air Force)

was powered by rocket engines, mounted on steel and aluminum "slippers" which slid on railroad-type rails. Convair engineers built a B-58 nose section complete with ejection systems. The nose section was mounted on the sled and shot down the 4.1-mile track. The seat, containing a 200-pound articulated dummy, would eject in the same fashion that a crew member would exit the B-58 in an emergency. The dummy was properly instrumented to determine what a human would be subjected to during an ejection. A parachute on the ejection seat brought the dummy slowly to the ground after its

B-58 SACseat ejection seat testing on 7 August 1958 using a 200-pound dummy indicated that the seat would not properly protect the crew member during supersonic flight. (National Museum of the U.S. Air Force)

successful ejection. Testing concluded that the SACseat would not provide adequate protection for the crew members when ejecting at supersonic speeds and Convair began studying other means of ejection.

A new escape capsule for each crew position was being developed by Stanley Aviation Corporation and progressing well, but it was experiencing marginal stability after ejection. The stabilization issue was eventually corrected through testing, and the modified capsules would self-stabilize after ejection. Retrofit after production was planned to start in late 1962. Each capsule required a four-point hookup (lap belt, torso restraint, oxygen hose, and interphone wire).[28] The capsules were very cramped and further reduced the size of the already small crew compartment. Crewmembers had to be "man-sized" to the capsule before beginning crew training. Failure to fit in the seat was cause to reject a would-be crewman.[29]

The capsules could eject using a rocket system from ground level and 120 knots to 70,000 feet and Mach 2.2. Each capsule was an independently operating unit which required no outside power source. The second and third crew stations were identical and were both ejected on vertical rails. The pilot's capsule was similar, but it encapsulated the flight control stick and included a window for the pilot to view the flight instruments. It was ejected on slightly leftward-canted rails. A three-piece telescoping clamshell door was pivoted on each side of the seat. It was stowed above the crew member's head during normal flight, like a big visor.[30]

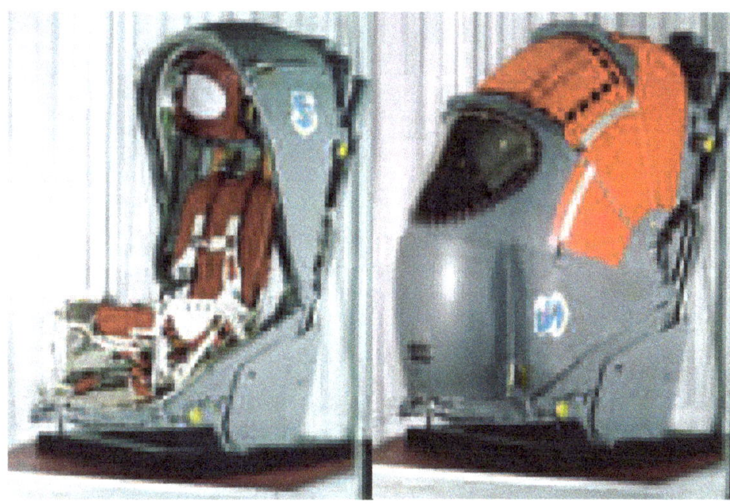

Convair B-58A ejection capsule. (U.S. Air Force)

Once the capsules were installed in the aircraft, operational flights were conducted in normal flight suits in a more or less "shirt sleeve" environment. Pressure-type oxygen equipment and protective flight helmets were worn.[31] The capsule doors remained open throughout the flight and the crew members had freedom of movement within their crew compartments. The compartments remained fully pressurized. However, the doors could be closed and the capsules separately pressurized in the event of an unplanned depressurization. The pilot could still access the control stick and fly the aircraft by viewing the instruments through the capsule door, pickling the autopilot using the release switch on the stick, manipulating the throttles to approximately 90 percent, shifting the CG with switches on the control stick, and descending to a safe altitude where the doors could be opened.

The ejection sequence was activated by raising the hand grips which activated the restraint harness system to rapidly pull the crewmember back into the seat. A horizontal bar would instantly pull the knees up into the chest and two retractor arms would pull the ankles in to ensure that the crewmember was entirely inside the capsule. The clam shell door then slammed shut in about a quarter of a second. Emergency oxygen and pressurization were automatically actuated, and each crew member ejected by squeezing one or both ejection trigger handles. This action fired the hatch jettison actuator and the rocket catapult initiator. The capsule ejected then stabilized and descended on a parachute. It had an independent pressurization and oxygen supply system as well as shock absorbers to ease the impact on touchdown.[32] Flotation devices were provided for water landings. The capsules also contained survival gear (clothing, radio, rations) that allowed crewmembers to survive up to three days without assistance.[33]

B-58 crew escape capsule set for initial high-speed testing on the sled at Edwards AFB, CA. (American Aviation Historical Society)

The escape capsules underwent initial sled testing from 1959 to 1961. High-speed testing to verify the operation of the stabilization chute was conducted at Edwards AFB, CA in which a rocket-powered sled slid down railroad-type rails at up to 794 mph. Next, the capsules underwent human factors testing at Holloman AFB, NM including deceleration tests. Finally, the capsules were subject to ejection testing at the Supersonic Military Air Research Track (SMART) in the Mojave Desert at Hurricane Mesa, UT. A nose section like the one used during SACseat testing was mounted to a sled which was powered by rocket engines and slid down a track using slippers like the SNORT at China Lake. The sled shot down the 12,000-foot course and the capsule was ejected as it neared the end of the track at the brink of a 1,500-foot cliff. The parachute then brought the capsule slowly to the ground on the valley floor where it was examined by engineers for damage. Over 50 sled tests were completed. Initially, the sled was set to travel down the track at subsonic speeds but eventually reached supersonic speeds as testing continued.

B-58 nose section on the rocket sled at the SMART facility in the Mohave Desert at Hurricane Mesa, UT. (American Aviation Historical Society)

The capsules were first static test fired at CVCFW on 5 and 12 April 1960. The first firing of a production prototype capsule with a production-type rocket catapult was completed in August 1960 at CVCFW. The capsule with a 200-pound dummy inside shot to a height of 200 feet and landed about 425 feet away in a circus net set up to prevent damage to the valuable capsule. The first static test of a capsule fired from a B-58 aircraft occurred in March 1961 from the navigator/bombardier's station of YRB-58A (55-0661) at CVCFW. The production-type capsule was weighted to simulate human occupancy and fired by remote control. The rocket catapult fired the green-colored capsule free from the aircraft and a split second later the rocket engine started, blasting the capsule about 250 feet up and forward. At its apogee, the parachute popped open and floated the capsule to a soft landing in a field 250 feet away. According to Convair, all four phases of the ejection – catapult, rocket engine, stabilization, and recovery – worked to near perfection.

Successful ejection capsule static test from the navigator/bombardier's station of YRB-58A (55-0661) in March 1961 at CVCFW. (National Museum of the U.S. Air Force)

On 27 May 1961, YRB-58A (55-0661) was assigned to the AFFTC at Edwards AFB, CA to support on-going capsule tests. A capsule ejection using an articulated dummy was conducted in October 1961 from the aircraft speeding down the Edwards runway at 115 mph. The capsule shot to 1,200 feet and landed to the left of the runway. This was followed in the same month by the first in-flight ejection at Edwards with the aircraft flying Mach 0.8 at 20,000 feet, as well as another Mach 0.8 and a Mach 0.9 ejection at 20,000 feet in December 1961. The first almost simultaneous ejection of two capsules also occurred in December 1961 at Hurricane Mesa, UT when the capsules were ejected one second apart from the rocket sled traveling 115 mph. The capsules containing dummies were ejected from the pilot position and then the navigator/bombardier position.

A chimpanzee was ejected over Edwards from the YRB-58A traveling at Mach 0.9 at 20,000 feet in early February 1962. The first manned subsonic capsule ejection

Edward J. Murray climbs from a capsule at Edwards AFB after the first manned ejection test from a Convair YRB-58A (55-0661) on 28 February 1962. (University of North Texas Libraries)

occurred on 28 February 1962 when ex-combat parachutist Chief Warrant Officer Edward J. Murray ejected while traveling Mach 0.8 at 20,000. Murray commented after his historic 7 minutes and 8 seconds descent, "No sweat – I feel fine. It was a much easier, nicer, and gentler ride than I anticipated. I can assure you it's a lot easier than open-seat ejection. I had no sensation of turning, spinning, or falling during the free fall before the chute opened." This paved the way for a series of supersonic ejections.

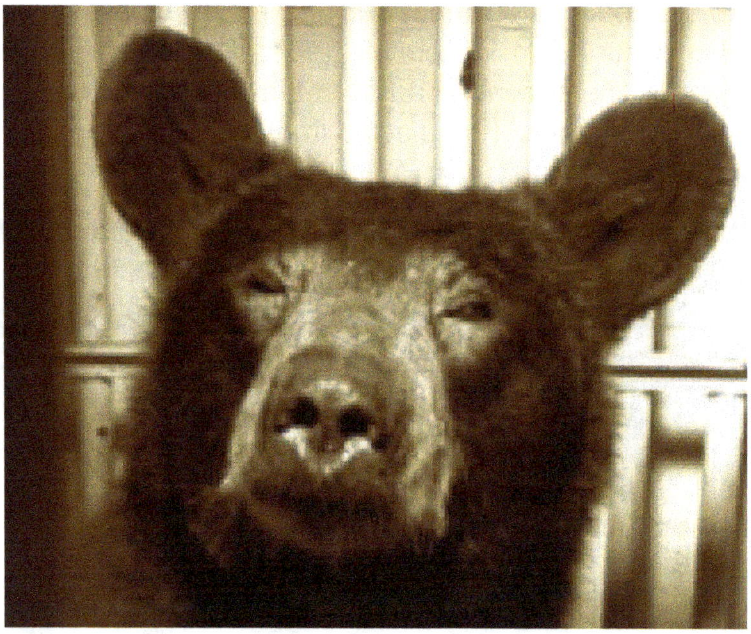

A black bear named "Yogi" became the first living being to survive supersonic ejection from an aircraft on 21 March 1962. (U.S. Air Force)

On 21 March 1962, a black bear named Yogi became the first living being to survive a supersonic ejection from an aircraft. The capsule was ejected at 35,000 feet from the YRB-58A (55-0661) flying at Mach 1.3 and the bear landed unharmed 7 minutes and 49 seconds later. Another bear named Big John was ejected flying at Mach 1.6 and 45,000 feet on 9 April 1962. The ejection proceeded as normal at first but the capsule soon experienced stability issues including more violent movements than normal and abruptly pitched up causing minor injuries to the bear's pelvic bone and neck.

A chimp was again ejected on 8 June at Mach 1.6 and 45,000 feet after appropriate repairs were made to correct the previous stability issues.[34] This test confirmed the repairs were successful even though the pitch rate initially increased sharply to 300 degrees per second but quickly decreased and remained low. The next ejection occurred on 20 June at Mach 0.8 and 40,000 feet with a bear on board once again. The test was adjusted to provide maximum pitch up to more thoroughly test the previous stability fixes.

Another highspeed runway ejection was completed on 22 June. This was similar to the runway ejection completed in October 1961. Ejection occurred at 115 mph with a bear in the capsule. The test was completely successful and verified the correct CG and rocket thrust relationship.

The first Mach 2.0 ejection, the fastest controlled ejection recorded, was conducted at 47,000 feet with a bear on board once again on 13 July 1962. A bear was also ejected at low-level altitude of 2,500 feet flying at Mach 0.9 on 27 July 1962.[35] Until this shot, all ejections had been between 20,000 and 47,000 feet at speeds ranging from subsonic to Mach 2.0. These tests uncovered some problems with the stabilizer and recovery chutes which were quickly corrected.

Escape capsule launched from DSO position. (U.S. Air Force)

The final flight in the series was another Mach 2.0 flight, this time at 42,500 feet, with a bear as the occupant conducted on 21 August 1962 at Edwards. According to C.C. Widaman, Convair's manager of the B-58 escape capsule program, "This last ejection was the most successful. In fact, it is the consensus of the test team that this was one of the best, if not the best ejection of the series."

Crews were trained using an In-Plant Crew Escape Capsule Simulator. This simulator allowed crews to experience "jumps" from three feet up in the CVCFW plant. It consisted of an operational escape capsule suspended from a frame about three feet above the floor. It was operated by an instructor whose control panel was mounted directly in front of the simulator. The crew member pulled the pre-ejection handle which activated an inertial reel to tighten the shoulder harness and pull the occupant back into the seat. Next, leg thrusters pulled the knees up and restrained the legs and feet well behind the capsule door. The instructor then manually closed the door to be sure it would clear the student, although in operation this would happen automatically.

After the occupant squeezed the ejection trigger, the capsule entered a simulated ejection. A sudden burst of air was introduced to simulate pressurization. The capsule shot up quickly and entered a windblast to simulate leaving the aircraft. The stabilization booms, which prevented pinwheeling in flight, flared back, and simultaneously the stabilizing parachute popped out (simulated). The capsule was then allowed to "free fall" below 16,000 feet where the main recovery chute opened, and flotation booms extended from the four corners of the capsule. The occupant then landed with a good simulation of an actual ejection. A Mobile Trainer Unit was later installed at Carswell to train 43rd BMW crew members.

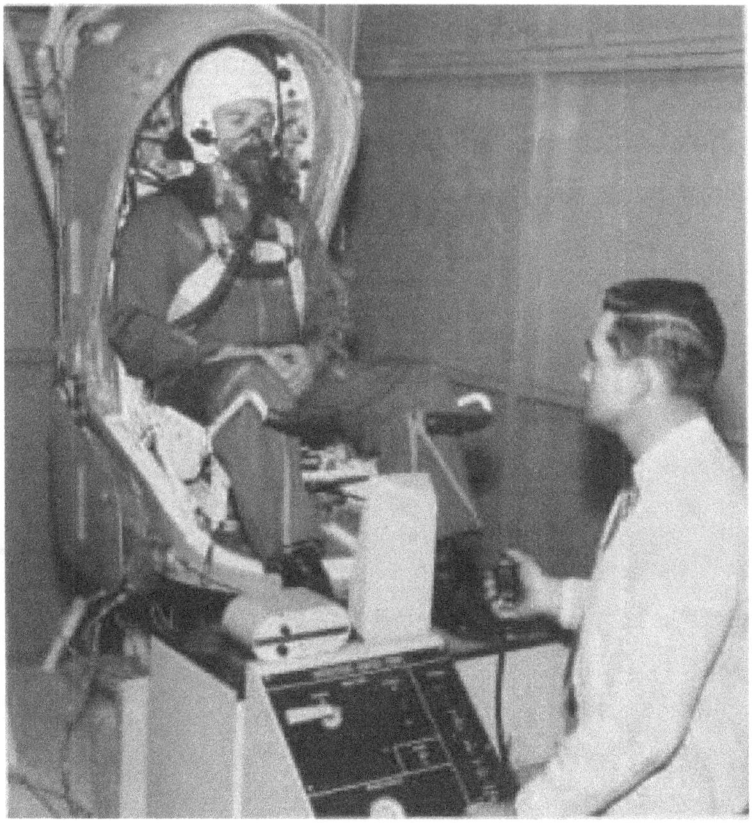

The B-58 In-Plant Crew Escape Capsule Trainer simulated the complete crew capsule ejection sequence to provide a realistic feeling of what it was like to eject. (General Dynamics News, 31 January 1962)

While the B-58's performance and design were exceptional for the era, some claimed it was not an easy aircraft to fly. This was caused by the 60-degree leading-edge sweepback of its wings and was inherent in these types of delta-wing platforms. It required a much higher AoA than a conventional aircraft, up to 9.4 degrees at Mach 0.5 and low altitudes. However, this problem was essentially eliminated as pilots gained more experience with the aircraft. Experienced pilots commented that the B-58 gave a smoother ride, had more responsive controls, and better stability than conventional aircraft. They typically flew the aircraft at Mach 0.91 cruise speeds and even low-level training missions never flew below Mach 0.7. The AoA at cruising altitude was typically about six degrees, regardless of gross weight, and about five degrees at low-level.

If the AoA was too high, in excess of 17 degrees, the bomber could pitch up and enter a spin. Several factors could prevent a successful recovery; if the pilot applied elevon, if the CG was not correctly positioned, or if the spin occurred below 15,000 feet, recovery might not be possible. The B-58 also possessed unconventional stall characteristics; if the nose was elevated, the bomber maintained forward motion without pitching down. Unless large amounts of power were applied, the descent rate increased rapidly. This issue was also essentially eliminated as pilots gained experience. There were no aircraft losses due to entering a spin.

The aircraft had unusual takeoff requirements, with a 14-degree AoA needed for rotation near 234.2 mph for a 150,000-pound take-off weight. It also had a high landing speed, necessitating a drogue parachute for braking. To accommodate the high landing speed, the specially configured landing gear had to handle excessive conditions; both the inflation pressure and wheel rotation speed were far greater than other aircraft. But these were normal characteristics for the B-58 and pilots soon became very skilled at performing safe take-offs and landings. At a 14-degree AoA, with the power of four engines in afterburner, the delta wing generated immediate lift and quickly reached take-off speed. Landings were made at high speeds because the B-58 did not have any flaps and a drag chute, like the B-47 and B-52, was used to slow it down. The landing gear and wheels were modified to withstand the high landing speeds.

Another problem pilots faced was called "fuel stacking", taking place whenever the B-58 accelerated or decelerated. It was caused by fuel movement within the tanks, which led to sudden changes in the aircraft's CG. This could cause the B-58 to abruptly pitch or bank, potentially resulting in loss of control. This issue was essentially non-existent in operational aircraft. The fuel system used in operational aircraft featured automatic in-flight pumping of the fuel to maintain the aircraft CG for various airspeeds. Manual fuel management responsibility was shared between the pilot and DSO.

A six-month study completed mid-1961 indicated several enhancements were needed to improve aircraft performance and bring all aircraft up to the final production configuration. Post-production modifications were initiated under the Project HUSTLE-UP two-phase program in CVCFW and San Antonio Air Materiel Area (SAAMA) at Kelly AFB, TX. Final approval of the project was given on 2 October 1962. B-58A (59-2449) arrived at CVCFW, which began operating under its parent company's name in September 1961 and would hereafter be known as General Dynamics Fort Worth (GDFW), on 24 October 1962 for repairs to its landing gear from damage incurred during landing in inclement weather. It consequently became the first aircraft entered into Project HUSTLE-UP Phase I on 30 October. Phase I involved 59 aircraft and included multi-weapons release capability, escape capsule installations, and voice warning system installations. The time required to complete the modifications varied by tail number but did not exceed 52 days. The first modified B-58A (59-2428) was completed and returned to the 43rd BMW at Carswell AFB, TX three days ahead of schedule on 18 January 1963. The final aircraft B-58A (59-2456) was completed in August 1963.

Phase II began in August 1963 and B-58A (59-2428) was the first aircraft inducted on 14 August. This phase involved 36 aircraft and 76 pods. It included reliability and maintainability improvements of the communications systems, flight control systems, and ECM. This phase took about 15 days per aircraft, and all modifications were completed two days ahead of schedule with the delivery of the final aircraft B-58A (60-1114) on 28

December 1963. The kits (aircraft, pods, training), and spares were acquired for $6.1 million. The capsule and weapons were the most extensive mods

Convair B-58A (60-1116) undergoing a major overhaul at SAAMA with several skin panels removed showing the internal aircraft structure. (National Museum of the U.S. Air Force)

The last B-58A (61-2080) completed its acceptance test fight on 23 October 1962 and was delivered to the Air Force, along with B-58A (61-2078 and 61-2079) on 26 October. Reliability was reaching new heights by the end of 1962. One B-58A (59-2454) assigned to the 43rd BMW achieved 31 consecutive on-time take-offs. This included four temporary duty sorties and meant the aircraft made all take-offs within 5 or 15 minutes[36] of scheduled time for nearly a year. This record reflected the improved maintenance performance and attested to the reliability that could be achieved with the aircraft.

But SAC initiated flight restrictions after three fatal accidents between March and September 1962 involving the flight control system. On 5 March, B-58A (59-2459) crashed on take-off from Carswell AFB, TX due to mechanical failure of the flight control system.

Last Convair B-58A (61-2080) delivered to the Air Force on 26 October 1962. (University of North Texas Libraries)

Reports said the aircraft lost one wing and went into a 6,000-foot skid. All three crew members perished. On 12 April, B-58A (59-2462) crashed shortly after take-off from Bunker Hill AFB, IN due to flight control system failure. The pilot and DSO ejected safely, but the navigator/bombardier was killed. On 14 September, B-58A (61-2057) crashed during normal flight three miles east of Butlerville, IN due to structural breakdown caused by flight control system failure. All three crew members perished.

Another major modification program centering on the flight control system was ordered. Phase I added a "gang bar" on the Yaw Damper switches along with other minimal improvements. Phase II modified the Mach Altitude Repeater and improved the unreliable amplifier computer circuitry, which allowed release of the flight restriction and resumed flights up to Mach 1.65. These modifications were redesignated as Phase I when they were completed in May 1963.

The remaining modifications planned for Phase II began on 1 April 1964 and included minor modifications of the flight control system, elevons, landing gears, and pylon sections of all aircraft. They were completed on Carswell aircraft by a Convair field team of 150 personnel in August. The aircraft assigned to Bunker Hill were completed by the same field team between September and December. The last 13 production aircraft of the 305th BMW were completed along with the remaining Project HUSTLE-UP modifications. Phase III was intended to improve the flight control system and remove

The Final B-58A (61-2080) rolled off the assembly line in October 1962 marking the end of B-58 production. (University of North Texas Libraries)

restrictions allowing Mach 2.0 flights. It was the costliest ($30 million) phase and not intended to start until the fall of 1966.

By 1963 the Hustler was proving itself a reliable and capable aircraft. General Power testified in front of the Preparedness Investigating Subcommittee of the Senate Armed Services Committee on 19 August 1963 and called it "one of the finest weapon systems in the world today." He continued, "I have particularly asked for B-58s. We have had it in the inventory for three years. We have had problems... but it was a big step forward in performance. We have more than doubled the speed of our bombers... It is a reliable weapon system, and I would like to see it put back into production."

But SECDEF Robert McNamara announced the B-58A phase-out in December 1965, despite continuing acceptance of the aircraft by Air Force leaders, which resulted in cancellation or delay of remaining improvement programs. Low-level flying capability was critical, but the TFR, under development and planned for implementation in 1967

($44 million), was cancelled in late 1965 leaving SAC with only a Radio Altimeter. BNS problems were well known, but the remedy was uncertain. Improvement proposals from various contractors were submitted in September 1965 but found unacceptable because they did not meet requirements, and the cost was estimated at approximately $70 million (twice anticipated). Even with the proposed changes many decision-makers doubted the system could become fully operational.

Efforts continued on an Inspect and Repair as Necessary (IRAN) program recommended by SAAMA in mid-1965. Problems were being fixed as needed but increasingly serious problems required attention including plumbing and wiring deterioration, signs of structural fatigue, and corrosion. SAC agreed and supported the 36-month schedule with the following caveats: fuel leaks were causing corrosion (more than estimated) and could not wait until the January 1966 implementation, and the 43rd BMW aircraft should be first. The IRAN initially began at GDFW and then moved to Connally AFB, TX (near

B-58A aircraft being reworked at GDFW. (University of North Texas Libraries)

Waco) on 1 July 1967. Major tasks were the removal of all "releasable" panels; inspection and repair of structures, wire bundles and cables, hydraulic lines and fittings, air conditioning and pressurization ducts. It also included bench check and calibration of all electronic units, landing gear overhaul, and treatment of corrosion. It required a total of 16,000 manhours per aircraft, which cost $181,000 in 1967 and $201,000 in 1968.

B-58A takes-off from CVCFW. (University of North Texas Libraries)

Some modifications were still completed despite the B-58A phase-out direction. The modifications were funded by mid-1968 after denial of the SAC waiver request for the 2-year utilization rule. Modifications were performed in GDFW and Waco. An improved Redundant Yaw Damper (RYD) was added to stabilize the erratic flight control system problems. It provided a redundant, gain scheduled, yaw damper with more authority, faster response, and higher reliability than the existing system. The RYD improved supersonic turn coordination, engine failure protection, dutch-roll damping, and automatic wing-heaviness correction. System flight testing started in May 1966 and included 193 flight hours over 110 flights. Kits were purchased in 1967 and installs started in May 1968. A new low-altitude Radar Altimeter was also installed to improve low-level flight capability. It provided both the pilot and navigator/bombardier with precise altitudes above ground level (AGL) with an accuracy of +/- 5 feet plus three percent of indicated altitude when flying below 5,000 feet.

Waco also began an overlapping modification program in June 1968 and completed 1 June 1969 to upgrade the BNS, install starter cartridges, and modify the flight control systems on 92 aircraft. AN/ASQ-42 BNS improvements, which replaced all 1950's era vacuum tube technology with very reliable solid-state components, were flight tested in mid-1967 and proved successful. These were subsequently approved on 27 September 1967. A cartridge self-starter (like the B-52) was pushed by SAC to eliminate the need for pneumatic ground carts for ground alert engine starts. The modification was approved, cancelled, reapproved, modified, and hampered by technical difficulties but it was finally approved, and installation began on 7 May 1968. The modification required the replacement of the J79-GE-5B engines with improved J79-GE-5C engines.

ECM improvements were also undertaken. Production systems were already obsolete, and the planned Mod 1180 was approved in mid-1966. This modification added a new ALQ-16 "trackbreaker" system. Other ECM improvements remained in work until the end of 1969 when the final decision that the B-58A would no longer be retained until 1974 resulted in cancellation of all penetration aid modifications.

Another IRAN contract (F41608-70-C-0002) was issued for FY70 despite pending aircraft retirements. The contract authorized general aircraft inspections, repair of structural defects, and various modifications. It covered 18 aircraft for $3.2 million with an option for a total of 36 aircraft at $6.5 million. However, all aircraft were retired by January 1970, and this contract was never carried out.

TB-58A. The TB-58A was a trainer version used to orient new pilots to the challenging flight characteristics and high-speed of the B-58A aircraft. SAC used the TB-58A to supplement pilot training by allowing new pilots to fly with an instructor before making solo flights in the B-58A. TB-58A aircraft supported three primary training requirements: transition and proficiency training for new B-58A pilots; instrument proficiency and procedures training; and standardization procedures training.

All TB-58A aircraft were converted from eight original YRB-58A R&D aircraft under contract AF33(600)-36200. Four aircraft were authorized in the first lot on 25 February 1959. Three of the four resulting TB-58A (55-0670, 55-0672, and 58-1007) aircraft were delivered to the 43rd BMW beginning in August 1960.[37] One TB-58A (55-0671) was delivered to the 305th BMW on 30 June 1961. An additional lot of four aircraft were authorized in another contract modification issued in May 1962. Three TB-58A (55-0661, 55-0662, and 55-0663) aircraft were delivered to the 305th BMW beginning in March 1963. TB-58A (55-0668) was delivered to the Air Force on 13 April 1964, and became the last aircraft delivered to the 43rd BMW on 14 April 1964. The total program cost was $16 million.

TB-58A aircraft were easily identified by the additional cockpit windows and no Vulcan cannon in the tail. The aircraft featured dual pilot seating. Most of the navigator/ bombardier's equipment was removed, his crew position was eliminated, and a new escape hatch was installed over a newly designed second station. All of the tactical equipment was removed and there was no autopilot, no primary navigation system, no bombing system, no defensive ECM system, and no active defense system.

TB-58A aircraft were easily identified by the additional cockpit windows and no Vulcan cannon in the tail. (U.S. Air Force)

Initially, the instructor pilot occupied the forward crew station, and the student pilot sat in the second station for preliminary indoctrination flights. The student then moved to the forward station to complete training and flight checkout. The second station (formerly the navigator/bombardier station) was modified with the necessary flight controls for in-flight instruction and safety. This included a set of conventional flight controls which were mechanically linked to those in the front cockpit. Every action the student took in the forward cockpit was duplicated in the second station. There was a control panel for the flight control systems in both stations, but the instructor pilot determined who maintained control. The second station also had control of fuel transfer operations. The instructor pilot's seat was elevated eight inches and offset to the right of center by

9.5 inches. From this position, 40 inches behind the cockpit, the instructor pilot could look through a glassed-in bulkhead into the cockpit to monitor the student's flight. The DSO (or a third pilot) manned the DSO position and assisted with co-pilot duties on these training flights. When the pilot was cleared to solo, the navigator/bombardier then joined the crew on his first mission in the B-58A.[38]

Oddly, the SACseat ejection seats used in the YRB-58A were retained in the TB-58A although they had been previously deemed unsafe. But this was necessary because the instructor pilot would have been unable to observe the student in the forward cockpit with capsules installed. The SACseats used a ballistic-initiated rocket-catapult ejection system which operated independently of other aircraft systems. The backrest of the seat provided storage for a back-pack type parachute, which crew members wore along with their helmets and oxygen masks during flight, and the bucket portion of the seat provided storage for the survival kit. A gas operated human-seat separation system facilitated separating the crew-member from the seat once clear of the aircraft.[39]

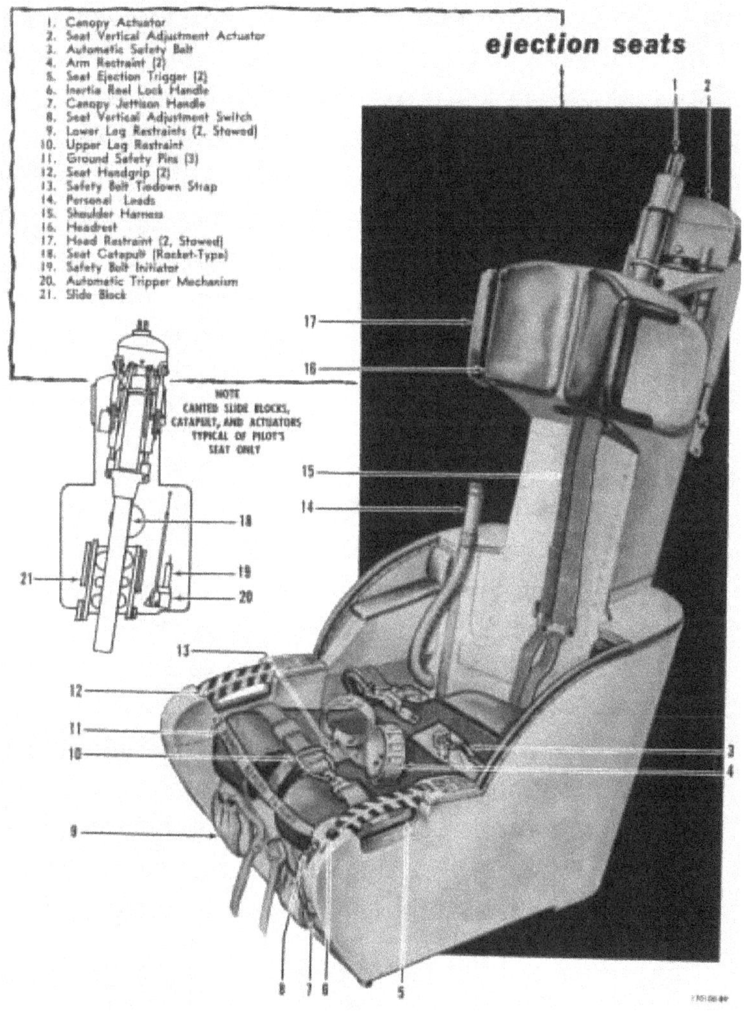

The Convair TB-58A retained the SACseat ejection seat to allow the instructor pilot to see into the forward station and observe the student pilot. (U.S. Air Force)

Each seat was mounted on ejection rails which were attached to each aft cabin bulkhead. Slide blocks on the back of each seat engaged the ejection rails and maintained the seat in a position such that its path of travel was parallel to the rails. The seat catapult, mounted on bulkhead attached brackets behind the seat, supported a seat adjustment actuator which in turn supported the seat and the canopy actuator. An initiator to blow the hatch was activated by a hand-operated trigger handle in each armrest and a second (delayed) initiator was used to activate the ejection seat rockets in the proper sequence for "safe" ejection. The pilot station had slightly angled rails to guide the seat out of the center of the hatch opening since the pilot seat was offset to the left for visibility around the center windscreen post.[40]

The first flight of the prototype TB-58A (55-0670) occurred on 10 May 1960. The aircraft was flown by Convair test pilot Val Prahl. The crew included Earl Guthrie in the instructor pilot position and Grover Tate in the third crew position. The aircraft flew without a pod and remained aloft for 1 hour and 40 minutes. It staged a supersonic dash and climbed to 44,000 feet. It flew most of the flight between 20,000 and 30,000 feet and went supersonic for about six minutes. The crew successfully demonstrated inflight operation of all systems peculiar to the TB-58A. Prahl and Guthrie passed flight controls back and forth several times during the flight demonstrating the ability to fly the aircraft from both the pilot and instructor pilot positions. Five more test flights were completed before the aircraft was turned over to the Air Force. The Air Force acceptance test, including supersonic segments, was completed on 26 July 1960.

Experiments & Tests

B-58B. Convair proposed a slightly enlarged and significantly more powerful B-58B, designated internally as B-58MI (Model Improved), as a follow-on to the B-58A. The B-58B would have been eight feet longer than the B-58A, with an extra structural section being inserted in the fuselage. One crew station was eliminated, and the two remaining were placed side-by-side. It was to be powered by four General Electric J79-GE-9 turbojets, which would have offered considerably more thrust than the J79-GE-5A and -5Bs of the B-58A. Gross weight was increased to 185,400 pounds and maximum fuel load was increased to 116,800 pounds. The B-58B was also to have featured some aerodynamic refinements, including canards, with extensions being added to the inboard leading edges of the delta wing, which would have improved the handling characteristics at high AoA. Wing area increased to 1,581 square feet and the wings were foldable. Jettisonable outboard nacelles were proposed to increase range performance.

Perhaps the most significant difference was in the B-58B being able to carry more weapons than the B-58A. In particular, the B-58B would have a significant conventional weapons capability by fitting a new and significantly larger under fuselage pod capable of carrying more fuel and conventional bombs in an internal bomb bay. In addition, a pair of air-launched ballistic missiles would be carried on underwing pylons just outboard of the centerline pod. Alternatively, an extra pair of fuel tanks could be carried on these underwing pylons.

The Air Force authorized development, design, and testing of the B-58B in October 1958. YRB-58A (55-0668) was designated as the B-58B testbed and was fitted with J79-GE-9 engines. A prototype B-58B (60-1109) was ordered and a total purchase of

Supersonic Transport (SST)

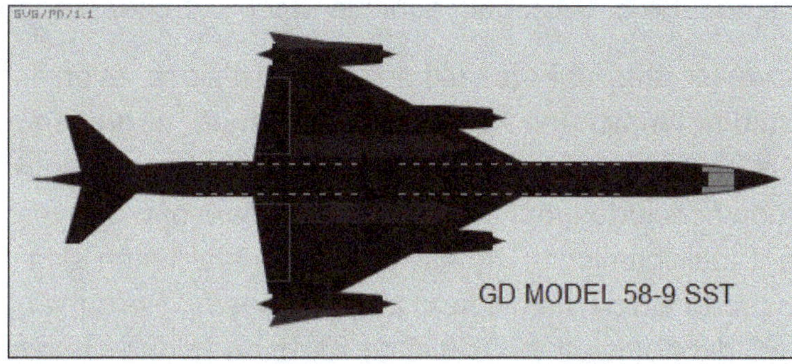

The Model 58-9 was an initial design of a commercial SST based on the B-58 capable of carrying 52 passengers. (Greg Goebel via VectorSite)

At an early stage in the Hustler project, there was some thought on the part of the Convair division of General Dynamics (GD) about producing a supersonic transport version of the B-58. Several different supersonic transport proposals were considered. Unfortunately, none of them actually got any farther than the initial concept stage. The simplest of these projects was a Convair proposal for a special B-58 capable of carrying a modified MB-1 pod that would accommodate five test personnel in comfort. The purpose of this special pod was to explore the problems caused by noise, heat, motion or psychological stress on various aspects of supersonic passenger flight. If this project had worked out, it is possible that a special passenger-carrying pod might have been built for the B-58 that would be capable of carrying a few government officials or Air Force officers at high speeds during times of national emergency. However, this project was scrubbed before it ever got past the initial concept stage.

In pursuit of a design that might appeal to the commercial airliner market, many initial concept studies were carried out by Convair during this period, with the final model being known by the company as Model 58-9. The Model 58-9 exhibited a clear family relationship to the B-58, but featured a long, slim 150-foot fuselage with a pressurized cabin that could carry a maximum of 52 passengers seated two abreast. The aircraft differed from the military B-58 in that the tailcone extended far behind the trailing edge of the delta wing and featured a separate horizontal tail. The aircraft was to be powered by four P&W J58 turbojets (two under the wings, and two on the wingtips), and could cruise at a speed of Mach 2.4 over a range of 2,900 miles. Gross take-off weight was expected to be 190,000 pounds. Various other seating arrangements were studied, including three- and four-abreast seating options. Unfortunately, the military showed very little interest in the project and the airlines were at that time undergoing a period of economic uncertainty, so the project was killed before any hardware could be produced.

(Excerpts from The B-58 Hustler Page)

185 aircraft including a production prototype B-58B (60-1110) was planned, but the B-58B project was cancelled on 7 July 1959 due to budgetary considerations before anything could be completed. The planned prototype B-58B (60-1109) was cancelled. The planned production prototype B-58B (60-1110) prototype was built as a B-58A instead.

B-58C. This was a proposed configuration for SAC use, designated BJ-58 by Convair, which was never produced. Significant airframe changes included a new wing leading edge, more tail area, and a 5-foot fuselage extension. The aircraft was planned to be equipped with four P&W J58 engines without afterburners. The J58 engines were the same as those used on the SR-71 and were capable of 18,000 pounds of thrust each (32,500 pounds with afterburner). The B-58C had an estimated top speed of Mach 2.4 at 68,000 feet, a super cruise of about Mach 2.0, a service ceiling of 70,000 feet, a maximum range of 6,000 miles unrefueled, and refueled range of 8,630 miles. Convair claimed the B-58C could achieve Mach 2.4 cheaper than the XB-70.

However, the Air Force did not pursue it since, even if it achieved the speed, the B-58C could not match the XB-70 payload, range, or growth potential. Also, the XB-70 design was just beginning whereas the B-58C was a continuation of an old design. An ARDC study concluded that the Convair estimate of 6,000 miles unrefueled range was 25 percent too optimistic. They also said the aluminum skin could create problems when exposed to aerodynamic heat at high speeds. Perhaps most important, the Air Force and SAC did not want the B-58C to compete for the remaining XB-70 development funds.

In early 1961, the new Kennedy administration canceled XB-70 production, reducing the program to only two experimental aircraft. In light of the loss of official interest in high-altitude, high-speed bombers, the Air Staff and SAC decided that they did not want to go through the trouble of obtaining funding for the B-58C project. In late April 1961 the Air Force told Convair there was no interest in the B-58C.

B-58D. In late 1951, Convair began developing proposals for a long-range interceptor and tactical variant of the B-58. In the fall of 1953, Convair submitted their proposal for a long-range all-weather interceptor to the Air Defense Command.[41] It was to be powered by two P&W J58 engines but was never produced. This configuration is often referred to as the B-58D although there is no clear documentation to substantiate this designation.

B-58E. During the summer of 1954, Convair presented its proposal for a long-range all-weather interceptor for TAC. It was to be powered by two P&W J58 engines but was never produced. Essentially the same as the B-58D. This configuration is often referred to as the B-58E although there is no clear documentation to substantiate this designation.

JB-58A. YRB-58A (55-0665) was initialy used by CVCFW for corporate service tests. It became the first YRB-58A formally delivered to the Air Force

JB/NB-58A (55-0665) was one of the 30 original YRB-58A aircraft used for service tests. It tested the AN/ASG-18 radar system and the GAR-9/AIM-47 missile, planned for the F-108 and YF-12A. (Air Force Flight Test Museum)

on 14 February 1958 before being transferred to the AFFTC at Edwards AFB, CA on 15 February for Category II testing under ARDC. It was modified in 1959 and redesignated as JB-58A on 12 September 1959 and then NB-58A on 30 November 1959 to test the Hughes AN/ASG-18 Fire Control Radar and the GAR-9/AIM-47 missile planned for the F-108 and YF-12A.

Modifications included a drooped and extended nose, dubbed "Snoopy", to accommodate the larger AN/ASG-18 radar dish, achieve aerodynamic efficiency, and allow adequate visibility for the pilot. The new nose added nearly seven feet to the overall aircraft length. An infrared sensor dome was mounted on each side of the forward fuselage to accommodate the AN/ASG-18 infrared capability. All BNS equipment was removed, and new controls and associated equipment were added in the navigator/bombardier station and other areas. A new pod was designed to carry and launch the GAR-9/AIM-47 missile and house the missile cooling equipment, radar components, telemetry equipment, and tracking flares. Another similar pod was fitted to another B-58 in August 1959 to support testing.

JB/NB-58A (55-0665) at Edwards AFB, CA. It had a modified "Snoopy" nose to accommodate tests of the Hughes AN/ASG-18 Fire Control Radar system and GAR-9/AIM-47 missile planned for the F-108 and YF-12A. (Air Force Flight Test Museum)

Flight testing began in early 1960 and the first missile launch occurred on 25 May 1962. By late 1963, testing was transferred to the YF-12 program and the last B-58 missile launch occurred in February 1964. The modifications were removed after tests, except the long nose, and the aircraft was towed to the photo range at Edwards for use as a ground trainer on 24 February 1964. It currently remains there in the "Museum Desert Storage Area" of the Air Force Flight Test Museum. The aircraft also flew a series of subsonic, low-level flights to measure response to turbulence at altitudes between 500 and 1,500 feet in 1961.

In addition, one production B-58A (61-2066) aircraft was initially delivered in place to GDFW before being transferred to the AFFTC for testing and redesignated JB-58A on 7 October 1963. It served as the primary testbed evaluating experimental, rubberized pneumatic pilot and crew seating systems. These were designed to dynamically inflate and adjust to mitigate severe fatigue, optimize blood flow, and improve overall crew comfort during prolonged supersonic missions. Because it retained its standard operational configuration while serving as a JB-58A, it was used to continuously map aerodynamic data across the B-58 Mach 2.0 envelope. It was assigned to 305th BMW as a SAC operational aircraft on 12 August 1964.

NB-58A. Four YRB-58A aircraft (including 55-0665) were converted for systems testing and temporarily redesignated as NB-58A. YRB-58A (55-0660) was the first B-58 aircraft built and first flew on 11 November 1956. It conducted the first launch of a missile from a B-58 flying out of Eglin AFB, FL on 5 September 1958 over the Cape Canaveral range. It was subsequently transferred to the 6592nd TS at Carswell AFB, TX on 24 December 1959 and redesignated NB-58A. It was returned to CVCFW on 24 February 1960 and redesignated YRB-58A. It was transferred to Kelly AFB, TX on 15 March 1960 for use as a ground trainer after completion of its test duties and later scrapped in-place.

YRB-58A (55-0662) was delivered in place to CVCFW and used for service test of the BNS and autopilot. It was also used for Doppler Radar and Radar Altimeter tests. It completed the first MB-1C pod drop on 5 June 1957. It was also the first aircraft to blow all tires on landing on 30 August 1958 and later used for frangible wheel testing. It was the first aircraft to complete Category I testing on 25 April 1959. It was then formally delivered to the Air Force on 29 April 1959 and retained at CVCFW. It was modified as the GE YJ93 engine test bed for the XB-70 and XF-108 beginning on 1 July 1959. It was then transferred from CVCFW to the AFFTC at Edwards AFB, CA on 24 September 1959 and redesignated NB-58A. The engine was installed in a special under fuselage pod. Several flight tests were completed including several flights at Mach 2.0. Eventually, the F-108 was cancelled, and the XB-70 was assigned to research only. The special pod was removed, and the aircraft was modified to the TB-58A configuration. It was temporarily returned to Edwards and used as the XB-70 chase plane before being assigned to the 305th BMW as a SAC operational aircraft. The aircraft set a record for 256 sorties without late or missed takeoff while assigned to the 305th.

Another YRB-58A (55-0668) was formally accepted by the Air Force on 13 May 1958 and delivered in place under bailment to CVCFW. It flew a total of 417.5 test hours over 126 flights. This included 108.5 hours over 40 flights for engineering development test

NB-58A (55-0662) was one of the 30 original YRB-58A aircraft used for service tests. It tested the GE YJ93 engine for the XB-70 and XF-108. (U.S. Air Force)

of the B-58 BNS. It was redesignated NB-58A on 13 January 1960 and used for testing the Hughes AN/APQ-69 Side-Looking Aperture Radar (SLAR), Goodyear AN/APS-73 Synthetic Aperture Radar (SAR), and other special projects devoted to the testing of advanced aerospace systems. It was also fitted with J79-GE-9 engines and intended to be used as the testbed for the B-58B. It was modified after the B-58B program was cancelled and redesignated TB-58A on 13 April 1964 for use as a SAC operational aircraft.

RB-58A. A photo-reconnaissance version of the B-58A, designated RB-58A, was an initial mission requirement. It was intended as a high-speed and high-altitude capability. On 27 May 1953, the Fairchild Camera and Instrument Company was given a letter contract for Phase I studies of a dedicated photo-reconnaissance pod for the B-58. These studies were completed in March 1954 and submitted for Air Force approval. The Air Force requested revisions to the specification on 20 July 1954 and approved the final version on 3 December 1954.

The specification required modifications to install the system in the standard MB-1C pod. The only airframe modifications would be replacement of the BNS equipment in the second crew station with the required photo-navigation equipment. The pod, designated MC-1, was to have a multi-camera system consisting of three 36-inch format cameras installed in a stabilized mount, a tri-camera system consisting of three six-inch cameras, one three-inch forward oblique camera, a camera control system, a nose-mounted television view finder, a fan of five three-inch cameras, a Melpar recording system, a Sperry navigation system, and a Raytheon search radar scope camera. However, in early July 1955, the reconnaissance pod was cancelled due to funding limitations. The program was reinstated in September 1955, when funding was again made available. The program was cancelled for good in early 1958, after only one pod was completed. This pod was never flown under a B-58, although one YRB-58A (55-0671) was scheduled as the testbed aircraft.[42]

In early 1952, the Air Force requested that Convair design and build an Electronic Intelligence (ELINT) pod. The MD-1 was a proposed electronic reconnaissance pod which used many of the shell components of the standard MB-1C free-fall pod. It was equipped with a wide range of electromagnetic energy sensors to analyze and record all enemy radar signals that reached the aircraft. A photographic system also allowed reference images of all questionable radar pulses. A recorder was provided to record and print out the navigation and weather data, as well as the corresponding Greenwich Mean Time (GMT). One pod was completed, but it was never flown.

In January of 1956, a special reconnaissance B-58 configuration was proposed which called for the development of a special pod that would carry a Hughes AN/APQ-69 SLAR. This project was officially started in September 1958 based on a Convair proposal for a special reconnaissance configuration submitted in January 1956. The pod and radar were built by Hughes Aircraft Company and delivered to Convair in February 1959. The system was tested through October 1959 before it was installed on YRB-58A (55-0668) and fit tested on 25 November. The first flight test took place on 24 December 1959. Some 25 test flights were completed through 30 June 1960. Results were considered "nominally good." The radar range achieved was about 50 miles with a resolution of 10 feet.

The AN/APQ-69 was a "real aperture" radar and required a 50-foot-long antenna making it one of the largest ever installed on an aircraft at the time. Because of the size of the antenna and the associated electronic systems, the pod could not carry any fuel.

This adversely affected the aircraft's range and limited mission time to 3 hours and 15 minutes. The pod's size and rectangular configuration (and its blunt nose) limited the aircraft to subsonic speeds when it was installed. It also presented problems with nose gear retraction because of the bow wave it created. In order to retract the gear, it was necessary to do a 0.5 g pushover shortly after takeoff. Consequently, further work on this radar reconnaissance system was abandoned.

In June of 1958, a more sophisticated B-58 reconnaissance proposal was issued by the Aerial Reconnaissance Laboratories of the Wright ADC. It was given the code name QUICK CHECK. The system was based on the Goodyear AN/APS-73 (XH-3) X- band SAR. This radar featured a simultaneous terrain mapping capability that could be used

First Invisible Super Hustler (FISH)

The B-58B was also planned to be the mothership for a highly classified Super Hustler parasite aircraft. The concept evolved quickly through several design configurations. However, reconnaissance soon became a primary focus for the Super Hustler. When it was formally presented as a preliminary study in September 1958, the aircraft was initially called the Special Purpose Super Hustler. This variant was a single-stage aircraft. It avoided detection from threat radars by operating at high speeds and altitudes. Radar cross section was addressed by the aircraft's small size and by its flat bottom. The gross weight was 31,200 pounds and cruise speed was Mach 4+, maximum altitude was 90,000 feet, and range was 4,500 nautical miles.

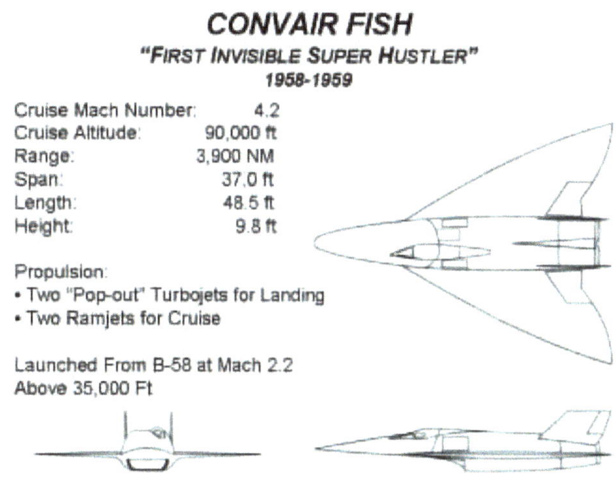

The Convair FISH was a conceptual design capable of Mach 4+ at a maximum altitude of 90,000 feet. (CIA)

The photo reconnaissance package offered two options. The first consisted of one large panoramic camera with a 24-inch focal length. At the aircraft's cruising altitude of 90,000 feet, the camera could cover an area of thirty miles by 1,500 miles while producing individual images measuring nine inches by 38 inches. The second option, called a dual-purpose photo reconnaissance installation, consisted of three smaller 2.25-inch square format cameras (three-inch focal length) designed to locate targets and one larger 4.25-inch by 18-inch panoramic camera (12-inch focal length) designed to collect technical intelligence. The signals intelligence package, called a ferret reconnaissance system, was designed to detect, locate, and analyze radar and radio station signals.

A month after the preliminary study was presented to the Air Force, Convair was asked to exploit the reconnaissance capabilities of the Super Hustler concept. This resulted in a Special Purpose Super Hustler Feasibility Study that was presented for initial evaluation on 11 November 1958. The study included two basic aircraft configurations. The first configuration had leading edges that were slightly curved outward to lower the radar cross section. The second had much larger curved wings and became the recommended configuration.

By December 1958, the design was 48.8 feet long and 8.7 feet tall. It had a gross weight of 35,000 pounds and an empty weight of 15,300 pounds. It soon became known as the FISH. The aircraft was powered by two Marquardt MA24E ramjet engines. A P&W JT12 series turbojet engine was nestled between and just in front of the ramjets. On 22 December 1958, Convair was given the green light to continue the development of the FISH.

The mission profile for the initial versions of the FISH involved launching the aircraft from the B-58B carrier at Mach 2.2. The FISH would then climb and accelerate to Mach 4+ in three minutes. In about another minute, the aircraft would reach its 90,000-foot cruising altitude. Convair continued to refine the FISH design over the first five months of 1959. The revised aircraft had several significant design changes. The single JT12 turbojet engine between the ramjets was replaced by two GE J85 turbojet engines located just behind the cockpit. The vertical tails were relocated from the wings to the fuselage. Gross weight increased to 38,325 pounds and range was reduced from 4,150 to 3,900 nautical miles.

Convair conducted hundreds of hours of wind tunnel testing and several balance, drag, and fuel problems surfaced. These problems could be addressed by increasing the length of the FISH by about 1.5 feet, the B-58 carrier by five feet, and by upgrading the B-58 to J79-9 engines. Convair proposed modifying two B-58s for the flight test phase of the program and two others as B-58B prototypes. Production B-58Bs would become FISH carriers. Consequently, when Convair failed to sell the Air Force on the B-58B carrier aircraft, the FISH was dead on the design table.

(Excerpts from Super Hustler, FISH, Kingfish, and Beyond by Eric Hehs, Code One Magzine, 16 September 2011)

on both sides of the aircraft at ranges of up to 80 nautical miles with a resolution of 50 feet. The QUICK CHECK project involved the modification of the aircraft's frame as well as the adaptation of a special MB-1C pod to carry the radar. Unlike the earlier AN/APQ-69 radar, the Goodyear AN/APS-73 required considerably less pod space, which permitted the pod to carry fuel. YRB-58A (55-0668) was modified to carry the QUICK CHECK pod in January 1960. A new and slightly bulged nose radome was fitted to accommodate a special Raytheon forward-looking radar, and the instrumentation installed at the second crew station was modified.[43]

The QUICK CHECK modified aircraft was delivered in May 1961. The Quick Check aircraft was subsequently used for an overflight of Cuba during the October 1962 missile crisis. The radar system functioned fairly well, but resolution was found to be the highest when subsonic speeds were used. Despite its promise, the QUICK CHECK program was cancelled following completion of the flight test program.[44]

A new program called Project MAIN LINE was established on 9 October 1963 based on results of the Cuban Missile Crisis. Several B-58A could be used for low-level/high-speed reconnaissance capability. Several options were rejected until the final solution was reached. It included a KA-56 panoramic camera and magazine in the MB-1C pod nose behind a V-shaped optical glass port along with a scanner system that pivoted the camera to provide unblurred horizon-to-horizon images at high speed and low level. The magazine held up to 1,000 feet of film. The control panel for the camera was mounted in the navigator/bombardier station. It was interchangeable with the weapon monitor and release panel for the bomb pod.

This option was approved by the Air Staff and implemented in two phases. Phase I provided an immediate photo reconnaissance capability and was successfully flight tested on 25 October 1963. Ten cameras were purchased, and ten MB-1C pods were modified along with 44 aircraft. Modifications were completed on 6 December 1963 at $1 million. Phase II was conducted by SAC personnel who installed modification kits on the ten MB-1C pods, which were given the designation LA-331. Phase II modifications were also incorporated into 24 aircraft by 30 June 1964.

The reconnaissance capability of the B-58 was employed on a number of occasions, mainly to provide quick photographs of natural disasters to enable a rapid assessment of the relief resources required. On 28 March 1964, the day after a major earthquake devastated Alaska, SAC tasked the 43rd BMW to undertake a special aerial reconnaissance mission to provide photographs of the region hit by the quake. Two B-58 aircraft were equipped with reconnaissance pods and took-off for Alaska within two hours after notification. When they arrived, the area was obscured by a layer of clouds at 1,000 feet. Mountains ringed the targets and the overcast made precision navigation critical.

The two bombers made five low-level runs over the devastated area between 300 and 400 feet AGL. The automatic cameras housed in the pods worked perfectly. The aircraft were dispatched to Offutt AFB, NE where the film was processed by the 544th Aerospace Reconnaissance Wing at SAC Headquarters. The Hustlers flew the 5,741 miles round trip to Alaska and back in 10 hours and 20 minutes. The next day two more Hustlers were dispatched for additional photo missions over the area. The crews of the four Hustlers were awarded the Air Medal in recognition of "acts of merit or sustained operational activities while participating in aerial flights." General Power cited the flights as "an outstanding example of SAC professional performance."

Armament and Weapons

Defensive Systems

Both active and passive defensive systems were controlled by the DSO. The DSO also monitored and transferred fuel to maintain proper CG throughout the flight (when not automatically maintained). The 20mm T-171E-3 (later designated M-61A1) Vulcan six-barrel rotary cannon in the tail section featured automatic operation after lock-on (displayed on the DSO screen) and had a firing rate of 4,000 rounds per minute.[1] It could fire-off 115 rounds in the time it takes to say "Gatling Gun."

Initially, Convair engineers faced a seemingly insurmountable problem with integrating the tail gun turret into the airframe. A bubble-type turret normally used for aircraft

Convair B-58A tail gun with 20mm T-171E-3 Vulcan six-barrel rotary cannon and conical turret composed of sectionalized, spring-operated plates allowing the turret to move in any direction. (University of North Texas Libraries)

weapons created turbulence in the airflow resulting in significant drag on a supersonic aircraft. They knew they needed a cone-shaped turret instead, but it would need to force its way into the airflow whenever pointed away from center. This would take tremendous power at high speeds and could act as a rudder affecting aircraft control. After some time, they designed a conical turret composed of sectionalized, spring-operated plates. The new turret would move in any direction with minimal airflow disturbance using minimal power. The firing zone was any target within a 30-degree cone.

The MD-7 FCS made the necessary calculations to aim the gun and instructed the DSO when it was time to open fire. The FCS consisted of a group of electronics packages which searched, acquired, and tracked targets; computed the fire control problem; and aimed the gun at the tracked predicted position. The FCS radar was a Ku-band system mounted directly above the gun behind a small bullet-shaped radome. Mach number and relative air density information were automatically supplied by the Air Data Computer. Gross weight was provided by the autopilot computer. A total of 1,200 rounds of ammunition were carried in a box in the fuselage just forward of the turret.

The FCS had four DSO-selectable operational modes; Search, Automatic Target Acquisition (ATA), Manual Target Acquisition (MTA), and Track. During Search mode, the FCS radar automatically searched a pattern at the rear of the aircraft and displayed target range and azimuth on the radar scope. In ATA mode, the FCS automatically detected radar targets, moved the gun to the computed firing position, armed the firing circuits, and then alerted the DSO with a time to fire indication. In MTA mode, the DSO manually identified the target through an extended lock-on range, and then all other functions were the same as the ATA mode. In Track mode, the FCS tracked and locked on a target and a time to fire indication was given to the DSO at the appropriate time.

Passive systems included an ECM system with an active jammer and chaff dispensers for disrupting enemy radar. The B-58 ECM system was the first production use of track-breaking where the range gate of the tracking radar was captured and led away from the attacker.[2] The system consisted of AN/ALR-12 Radar Warning equipment and AN/ALQ-16 Radar Track Breaking equipment. YRB-58A aircraft also had AN/ALQ-15 Radar Confusion equipment, but this was removed in favor of the AN/ALQ-16 for production B-58A aircraft. An AN/ALE-16 Chaff Dispensing System was installed in each upper main gear fairing, with chaff being ejected through mechanically actuated slots in the tops of each wing fairing.

The AN/ALQ-16 was a repeater type jammer that generated deceptive radar jamming signals as a function of RF energy received from enemy tracking radars. When tracking radar signals were received, the track breaker generated and transmitted deceptive angle and range information, causing the tracking radar servo system to generate false antenna positioning information which in turn caused the tracking radar to compute false range information. The equipment for each track breaker system consisted of receiver-type antennas, transmitter type antennas, a receiver-locked oscillator, a driver amplifier, a power amplifier, and a solenoid power supply.[3]

Offensive Systems

The B-58 Hustler's relatively small and narrow fuselage meant the internal fuel load was limited and it was unable to carry weapons internally. Convair solved this problem by using underbelly pods to carry additional fuel and a nuclear weapon. Two types of pods were available.

MB-1C Pod. The MB-1C pod was carried at Mach 0.91 cruising speed and released on a Mach 2.0 dash over the target at 55,000 feet. It was a standard free-falling unit that was carried under the B-58 fuselage on the centerline. It was basically a finned aerodynamic shell with a pair of fuel tanks carrying 4,172 gallons of fuel and a W39Y-1 variable-yield nuclear warhead. The pod was 57 feet long with a diameter of 5 feet. Empty weight was 8,550 pounds (including ballast), but when fully loaded with fuel and warhead it weighed 36,087 pounds.

Convair YRB-58A (55-0660) with MB-1C fuel and weapons pod. (U.S. Air Force)

The pod was attached to the aircraft by three pneumatically actuated hooks (one forward and two aft) which were mounted to the pod. A pylon enclosed the release system and components of other pod systems installed on the upper surface of the pod. An electrical disconnect mechanically connected electrical circuits of the pod to the aircraft. Fuel and fuel pressurization disconnects engaged matching components on the aircraft. These disconnects allowed the pod and aircraft fuel systems to be connected and pressurized air to enter the fuel system from the aircraft. All disconnects were released and closed instantly when the pod was released from the aircraft.

The pod had an equipment bay, a forward fuel tank, a munitions bay, an aft fuel tank, a tail cone and fins, and an attachment pylon. The equipment bay extended from the nose of the pod to the forward fuel tank bulkhead. The forward fuel tank and munitions bay were divided into two sections by the munitions loading splice. The loading splice allowed installation of ballast in the munitions bay. The four fins were mounted at 45 degrees from the horizontal centerline and were slightly offset to give the pod a slow rotation during its trajectory. The warhead was fused by a set of barometric switches, set to trigger the weapon when the preset pressure was reached.

Crew loads an MB-1C pod on a YRB-58A in 1958. (University of North Texas Libraries)

MB-1C pod drop testing occurred in several phases at Holloman AFB and Kirtland AFB, NM. The pods, many heavily instrumented and equipped with radios for data transmission, were trucked from CVCFW to Kirtland on a fixed schedule. The first YRB-58A used to support MB-1C pod drop testing arrived at Holloman AFB, NM on 26 November 1956. On 16 February 1957, YRB-58A (55-0661) became the first to fly with a test MB-1C free-fall pod attached reaching Mach 1.15. On 5 June 1957, YRB-58A (55-0662) made the first MB-1C pod drop while flying Mach 0.9 at 40,000 feet over the Holloman AFB test range. On 29 June, YRB-58A (55-0660) reached Mach 2.03 at 43,350 feet carrying a dry MB-1C pod becoming the first aircraft to reach Mach 2.0. The flight lasted 1 hour and 55 minutes.

Successful drops took place at progressively higher speeds, culminating when YRB-58A (55-0663) made the first supersonic MB-1C pod drop while flying Mach 1.4 at 40,000 feet on 30 September 1957. A pod drop at Mach 1.6 subsequently occurred in October and a Mach 1.8 drop occurred in November. This was followed by two successful pods drops from 42,000 feet at Mach 2.0. Finally, YRB-58A (55-0663) completed a pod drop at Mach 2.0 from above 60,000 feet for the first time on 20 December 1957. This aircraft made a total of 10 MB-1C pod drops.

MB-1C pod testing continued throughout 1960. YRB-58A (58-1011) became the first to drop a pod using a fully functional BNS on 12 February 1960. The first drop of an MB-1C pod with an inert TX-53 nuclear warhead installed occurred in May 1960. YRB-58A (55-0672) also participated in pod drop testing while deployed to Kirtland AFB, NM.

BLU-2B Pod. Problems with fuel leaks led to the development of a new Two-Component (TC) pod, the BLU-2B. The TC Pod was comprised of two pods, an upper BLU-2B-3 bomb pod, and a lower BLU-2B-2 fuel pod wherein the upper section was mostly inserted into the lower section using a crane. The lower fuel pod was expendable and was jettisoned during flight when all the fuel in both the upper and lower components was consumed. The upper bomb pod remained with the aircraft for release during the bomb run. This allowed the pilot to "clean-up" the aircraft for fuel efficiency or during emergencies while still retaining the weapon.

The upper pod was 35 feet long with a maximum diameter of 3.5 feet. It consisted of a forward fuel tank, a munitions bay that accommodated a megaton range BA53 nuclear weapon,[4] an aft fuel tank, a tail cone and three fins, parachute retardation system, plus an attachment pylon. Two of the fins were mounted on the sides of the pod at 30 degrees above the horizontal, and the third (lower) fin was retracted within the upper pod while the lower pod was still attached. It deployed automatically upon lower pod release. The empty weight was 9,540 pounds, whereas the gross weight with maximum fuel and a BA53 bomb was 11,970 pounds. A ballast was installed in the munitions bay when no weapon was installed.

The upper pod was carried beneath the aircraft by three pneumatically actuated hooks (one forward and two aft) which were mounted to the pod. The two aft hooks were

Convair YRB-58A (58-1011) with MC-1B pod (installed) and TC pod (on trailer). (U.S. Air Force)

mechanically connected to each other and to the hook actuator by a torque tube. The forward hook was linked to the torque tube by the pod release actuator tube. The pod electrical circuit was connected to the aircraft electrical circuits through the pod electrical disconnect. Pod stabilization after release was achieved by the three fins. A pylon contained the release system and various components of other pod systems installed on the upper surface of the upper pod. All disconnects released and closed simultaneously when the pod was released.

The lower pod was 54 feet long with a maximum diameter of 5 feet. It was divided into two tanks, called forward and aft fuel areas, separated by a common bulkhead and capable of carrying 3,885-gallons of fuel. It was supported underneath the upper pod by one forward and one aft releaser. There were no fins, but a pivot strut was mounted

(Right) The upper section of the TC pod is lifted from its trailer using a crane and loading beam. (Center) The upper section is lowered into the lower section where the two components are securely locked together. (Left) The two components are then ready for mounting to the aircraft in the same manner as the MB-1C pod. (Convairiety Newsletter, 9 November 1960)

on the aft end of the pod to facilitate proper separation from the aircraft during release. Fuel and fuel tank pressurization disconnects engaged the matching components on the upper pod and aircraft. All disconnects were released and closed instantly when the pod was released. The pod empty weight was 1,900 pounds, whereas the gross weight with maximum fuel was 26,000 pounds.

B-58A (59-2458) at an airshow with the BLU-2B TC pod installed. (American Aviation Historical Society)

Separation and accuracy tests of the TC pod at supersonic speeds were completed in April 1960. Nine runs at Mach 1.3 to 1.68 were conducted. The TC pod was first flight-tested on 12 May 1960, and the first drop of the lower component occurred on 24 May. The following major milestones were achieved throughout the pod drop program using YRB-58A (55-0663) and B-58A (59-2435). YRB-58A (58-1015) flew TC pod performance and stability demonstrations at Edwards AFB, CA on 10 April 1961.

- B-58A (59-2435) made the first upper component drop White Sands Missile Range in October 1960.
- YRB-58A (55-0663) made the first low level drop of the TC pod upper component at Tonopah Test Range, NV on 19 November 1960.
- YRB-58A (55-0663) made the first supersonic upper TC pod drop on 11 December 1960 flying Mach 1.6 at 45,000 feet over White Sands Missile Range and continued supporting pod testing by dropping TC pod upper components and seven lower components (several at low altitudes).
- B-58A (59-2435) made the first Mach 2.0 drop of an upper component on 10 February 1961.

- YRB-58A (55-0663) completed the first drop of both upper and lower components at Mach 2.0 to verify CG and drag abnormalities in April 1961.
- B-58A (59-2435) made two drops of the TC pod (dropping the lower component before the upper) and one drop of the upper component only in May 1961.
- YRB-58A (55-0663) performed low-level TC pod drops over Tonopah Test Range, NV in June 1961 to test the pod retardation system.
- YRB-58A (55-0663) made the first low-level TC pod upper component drop using an Air Force crew at the Tonopah Test Range, NV in August 1961. The same crew dropped the complete TC pod, with both upper and lower components, at high level over the Holloman AFB, NM range using B-58A (59-2435) a week later.
- YRB-58A (55-0663) performed a low-level drop at Eglin AFB, FL in August 1961 to test the retardation system at sea-level (also, high-level to test water impact).
- B-58A (59-2435) made the first Mach 2.0 lower component drop on 8 August 1961.
- B-58A (59-2435) made a successful supersonic pod drop over water at Mach 1.6 and 35,000 feet near Eglin AFB, FL in September 1961.
- B-58A (59-2435) completed pod drop testing at Kirtland AFB, NM with two high altitude TC pod drops in March 1963 followed by two more drops over the Eglin Gulf Test Range.

Multi-weapons Capability. The B-58 also had a multi-weapons capability. Hard points for four 2,100-pound, megaton range, Mk-43 nuclear weapons were added between 1962 and 1963. Capability was subsequently added to carry four kiloton range B-61 bombs. The weapons were mounted on external pylons underneath the wings between the fuselage and the main landing gear bays. Two weapons were carried on either side of the fuselage in tandem. Each weapon was attached to a hard point rack by cartridge actuated hooks (one forward and one aft) which latched on lugs attached to the top of the weapon. The racks contained the release mechanism and electrical disconnects for monitoring and were enclosed in streamlined pylons.

The B-58 program office and SAC began exploring the possibility of carrying multiple weapons on pylons on 27 September 1960. This evolved into a documented SAC operational requirement and recommendation for a Phase I study in December 1960. Phase I testing was authorized in January 1961 to prove the feasibility and compatibility of the B-58 to carry the added weapons. Preliminary testing was conducted on B-58A (59-2456) which was delivered in place to CVCFW under bailment on 6 February 1961. These preliminary tests were successful and resulted on 10 March in approval of the capability for B-58A (61-2051) and on. Flight tests to check out aerodynamic and ballistic reactions to flight were launched in June 1961 using two B-58A (59-2435 and 59-2439) aircraft. B-58A (59-2439) subsequently underwent ground suspended vibration tests and flutter analysis in November 1961.

B-58A (59-2435), named Shackbuster, made the first multi-weapons drop in February 1962 and the first supersonic multi-weapons drop on 19 June 1962. (University of North Texas Libraries)

Phase II flight demonstrations were conducted again using the same two B-58A (59-2435 and 59-2439) aircraft beginning in January 1962. B-58A (59-2439) was used for multi-weapons drop testing at Kirtland AFB, NM and flew captive flights from GDFW to determine performance characteristics of the new configuration. B-58A (59-2435), named Shackbuster, made the first multi-weapons drop on 7 February 1962 over White Sands Missile Range, NM. By May, the aircraft had completed five drops at subsonic speeds at altitudes from 36,000 to 40,000 feet and one from about 1,000 feet.[5] It made the first supersonic multi-weapon drop on 19 June 1962. The same aircraft made the first Mach 2.0 multi-weapon drop on 2 August 1962.

Mk-43 Gravity Bomb. The Mk-43 was 18 inches in diameter and 150 to 164 inches long. It weighed 2,060 to 2,125 pounds and produced a yield in the megaton range. The main components of the weapon were a jettisonable nose cone which covered the

Convair B-58A (59-2456) in flight with multiple weapons loaded during testing April 1961. (University of North Texas Libraries)

impact spike, a center section which housed the warhead, and a tail section which housed the retardation parachute and supported four tail fins. It could be dropped free-fall or retarded with parachutes and detonated with either air-burst, surface contact, or laydown. The Mk-43 was produced from April 1961 to October 1965. Four Mk-43 bombs could be carried on pylons attached to hardpoints under the B-58 wings.

Four Mk-43 bombs each with a yield in the megaton range could be carried on pylons attached to hardpoints under the B-58 wings. (U.S. Air Force)

Mk-53 Gravity Bomb. The Mk-53, redesignated B-53 in 1968, was 50 inches in diameter and 148 to 150 inches long. It weighed 8,850 to 8,900 pounds and produced a yield in the megaton range. It could be dropped free-fall or retarded with parachutes and detonated with either airburst, surface contact, or laydown. The Mk-53 was produced from August 1962 to June 1965. The Mk-53 was retired from operational use between July 1967 and early 1997. The Mk-53 was replaced by the B-61 and B-83 bombs. One BA53 nuclear bomb, a smaller version of the Mk-53 with no fins, could be carried in the munitions bay of the upper TC pod.

One BA53 nuclear bomb could be carried in the munitions bay of the upper TC pod. (National Museum of the U.S. Air Force)

B-61 Gravity Bomb. The B-61 was a lightweight multipurpose tactical and strategic weapon. It was 13.3 inches in diameter and 141.64 inches long. It weighed 695 to 716 pounds and produced a yield in the kiloton range. It could be dropped free-fall or retarded and detonated with either airburst or laydown. The B-61 was produced from October 1966 to the early 1990s. Early versions of the B-61 were retired from operational use during the 1970s and 1980s. It had the longest production run of any U.S. nuclear weapon and is the oldest design still in service. It is also part of the U.S. enduring stockpile. Four B-61 bombs could be carried on pylons attached to hardpoints under the B-58 wings.

Other Capabilities. To increase the flexibility of the B-58 as a weapons system, experiments were carried out in April 1964 under a program known as Operation BULLSEYE to see if the B-58 could carry and deliver conventional bombs. In coordination with Republic F-105D and McDonnell F-4C/D aircraft, sorties were flown in formation on low-level strike test missions out of Eglin AFB, FL using a B-58A (60-1119) as the lead ship and pathfinder and as an independent strike aircraft.[6] In one 27-day period about 75 sorties were flown. These flights demonstrated that the B-58 could carry iron bombs on pylons attached to hardpoints under the B-58 wings that had earlier been added to accommodate four nuclear weapons. Iron bombs of varying weights up to 3,000 pounds were dropped, usually from low altitudes and at speeds of 690 mph. Almost all of the drops were visual, with the AN/ASQ-42 BNS rarely being used. The disadvantages to the formation approach included difficulty in maintaining visual contact in bad weather, susceptibility to SAM activity when flying in the required tight formation, and an only marginal improvement in bombing accuracy. Furthermore, the fear that the B-58's

Convair B-58A (59-2456) demonstrates multi-weapons capability. (U.S. Air Force)

integral wing tanks would make it vulnerable to ground fire during low-altitude delivery led to the abandonment of the program.[7]

Still, the Air Force considered using some B-58 aircraft for a conventional role in Vietnam as lead aircraft with four F-105s flying in formation with the Hustler. The tactic would be for the B-58 to navigate to the target at night and at low altitude. All aircraft would drop their bombs on command of the B-58. Also, there would be mutual electronic jamming of enemy radars by all five aircraft. The actual flight testing of this concept with both F-4 and F-105 aircraft proved that it was do-able and would be very effective. However, the B-58 was ultimately not approved for use in Vietnam. The primary reason was because the B-58 was too valuable an asset for its role in the Single Integrated Operational Plan (SIOP) and therefore should not be used in a conventional role.[8]

The MA-1C pod was a proposed rocket-propelled air-to-surface missile version of the MB-1C free-fall pod, designed to give the B-58 a stand-off capability. The pod was powered by a Bell Aerospace LR81-BA-1 rocket engine, fueled by a combination of JP-4 and red fuming nitric acid. The maximum range was expected to be 160 miles. During the flight to target, a maximum altitude of 108,000 feet and a maximum speed of Mach 4 was to be obtained. Total powered flight time was 65 seconds. A Sperry guidance system was to control the pod during its flight to the target. Before launching, the guidance system determined the pod's present position and computed the desired heading.

After launching, it controlled the operation of the rocket engine. It also ordered the pitch angle for the climb and glide and dive angles to the target.[9] However, it was not accurate enough and the MA-1C pod was cancelled before it could be deployed after $66.4 million in development costs.

Several proposals were considered in spring 1958 for using the B-58 as a strategic missile launcher. These proposals gained significant interest, and the Air Force agreed to fund an experiment. Lockheed Missiles and Space Systems was given a contract to develop an Air Launched Ballistic Missile (ALBM) under project number ALBM-199C which also carried the code name HIGH VIRGO. The missile was developed using proven components from existing missile programs including the XQ-5, X-17, Polaris, and Sergeant. It was 30.3 feet long, 31 inches in diameter, and weighed about 12,000 pounds. It was powered by a Thiokol TX-20 solid-fuel rocket motor with a thrust of 50,000 pounds for 29 seconds.

YRB-58A (55-0660) conducted the first launch of an ALBM on 5 September 1958. (American Aviation Historical Society)

A YRB-58A (55-0660) flying out of Eglin AFB, FL conducted the first launch on 5 September 1958 over the Cape Canaveral range. Unfortunately, the missile failed and fell into the Atlantic Ocean 30 seconds later. The second missile was launched from the same aircraft flying at Mach 1.6 on 19 December and this time the results were much better. The missile achieved Mach 6 and total range was 185 miles at a maximum altitude of 263,000 feet. A third launch, on 4 June 1959, was also successful and a maximum altitude of 32 miles was achieved.

In the fourth and last launch, on 22 September 1959, the ALBM was programmed to photograph the Explorer V satellite under Project SNAPSHOT. This was a proof-of-concept test with applications for satellite inspection and anti-satellite interception. The missile was launched from the aircraft flying at 37,500 feet and Mach 2.0. The missile launched smoothly, but 30 seconds later all contact was lost. The B-58 ALBM program soon lost interest and was then abandoned.

Organization and Basing

At the time of the Japanese surrender from World War II, the AAF had 218 operational groups. By January 1946, the United States was rapidly demobilizing its military forces and the AAF was down to 30 percent of its wartime strength. The War Department's postwar planning called for a "bedrock minimum" air force of 70 groups and a personnel strength of 400,000. General Carl A. "Tooey" Spaatz took over as Chief of Staff in February. He held to the objective of 70 air groups and was supported in that position by the Army Chief of Staff, General of the Army Dwight D. Eisenhower. When Spaatz established SAC on 21 March 1946, it inherited a force of approximately 1,300 bomber, fighter, reconnaissance, and support aircraft. The initial SAC force included 100,000 personnel with 22 major installations and 30 minor bases.

However, the 70-group plan evaporated in August 1946 when President Truman ordered "economy commitments" to reduce the budget by $2.2 billion, with 75 percent of that to come from the armed forces. At the end of 1946, the AAF had 55 groups remaining, only two of which could be counted as combat ready. This would soon be reduced to 48 groups and a personnel strength of 304,000. SAC was reduced after the demobilization to a force of 279 aircraft including 148 B-29, 85 P-51, 31 F-2, and five C-54. Personnel were reduced to 37,092 with only 18 bases remaining.

SAC's bomber force was organized into nine Bomb Groups (BG) comprised of B-29 bombers. Three of these groups had no aircraft assigned. Each BG was an operational combat unit typically comprised of up to three operational squadrons assigned directly to the group headquarters. A base commander, often a non-flying administrator, was the highest echelon of command on the base and controlled all support functions. The base commander was also the immediate superior of the group commander.

Beginning on 17 November 1947, the 7th BMW was formed at Carswell AFB, TX as a test unit for the newly developed "Wing-Base" plan, also known as the "Hobson Plan", in which the wing became the predominant operational unit on each base. The wing commander, a highly experienced aviator, became the operational commander and the combat group reported directly to the wing. The combat group was still comprised of up to three operational squadrons and a group headquarters responsible for all flying activities. The wing commander also controlled three support groups, consisting of maintenance and supply, airdrome, and medical, which gave the wing complete control of all flying and support functions. The base commander now had responsibility only for base housekeeping and administrative functions.

The 11th BMW at Carswell was established in 1951 as part of a reorganization of SAC's combat command structure, based on a plan devised by General LeMay, using experience gained from Korean War operations. LeMay wanted an organization designed for peacetime that required minimum change to rapidly switch to war footing. Previously,

B-36 aircraft of the 7th BMW on the tarmac at Carswell AFB, TX. (American Aviation Historical Society)

despite having a base commander assigned, the wing commander was still occupied with too many base housekeeping and administrative tasks that distracted from his ability to focus on combat operations. Also, the wing commander relied heavily on the combat group commanders to conduct daily flying operations. With the reorganization, the Air Base Group commander assumed all base housekeeping and administrative duties. Although the combat groups continued to exist, they became subordinates to the wing and the wing commander served as the group commander. This gave full control and focus of combat operations to the wing commander.

In conjunction with this reorganization, SAC received authority from Headquarters Air Force to organize Air Division (AD) headquarters on double wing bases and to operate only one Air Base Group on these installations. The AD commander, typically a 1- or 2-star general officer, exercised direct control over the two wing commanders and the Air Base Group commander. SAC fully implemented this change across the command over the following year. On 16 June 1952, all SAC combat groups were eliminated, and the combat squadrons reported directly to the wing.

All SAC wings and ADs reported directly to a Numbered Air Force (NAF) commander. SAC employed up to three NAFs during its history, the 2nd, 8th, and 15th. Initially, medium and heavy bomber units were assigned to 8th AF while 15th AF had mostly medium bombers and 2nd AF concentrated on reconnaissance. This arrangement resulted in an unbalanced organization with wings in the same geographical region being assigned to different NAFs. On 1 April 1950, SAC reorganized the NAFs on a geographical basis with 2nd AF in the eastern, 8th AF in the central, and 15th AF in the western United States.

When former General of the Army Dwight D. Eisenhower became President of the United States in 1953, he assembled his advisors for discussions regarding defense strategy. The U.S. had just finished fighting a war in Korea that ultimately became a stalemate. Most observers believed the Soviets instigated it in their quest to spread Communism across the world. Eisenhower believed the Soviets could provoke more confrontations when and where in the world they wanted. The U.S. was unable to match the Soviets in conventional forces due to its drawdown after WWII. The Soviets could draw on at least 175 regular and 125 reserve divisions, compared to only 29 U.S. regular and seven reserve divisions.

Eisenhower soon decided on a "New Look" defense strategy that was outlined in NSC 162/2, dated October 1953. The new strategy relied on nuclear weapons to provide a deterrence against Communist expansion. It was designed to provide a massive retaliatory strike in the event of unprovoked attacks. Eisenhower saw this strategy as the most economical way to deter Communist aggression and protect the U.S. homeland. While the U.S. was considerably behind the Soviets in conventional might, it was significantly ahead in atomic stockpiles with 841 weapons compared to only 120 for the Soviets at the end of 1952.

This new strategy had a significant impact on the Air Force, which was given the primary nuclear delivery role for the U.S. military. The B-36 was in service and capable of carrying considerable nuclear payloads. The B-47 was entering service, and the B-52 was planned to begin operational service in 1955. The Air Force budget was about 47 percent of the DoD appropriation compared to 29 percent for the Navy and 22 percent for the Army. Air Force combat strength grew from 95 wings in the last Truman budget to 137 wings in the first Eisenhower budget. SAC was assigned a total of 92 wings and quickly equipped them with hundreds of B-47 bombers.

B-47 bombers deployed to Sidi Slimane AB, Morocco on 19 February 1954. (American Aviation Historical Society)

In 1955, SAC began converting all B-36 wings and some B-47 wings to B-52 aircraft. The Air Force Council recommended a B-52 fleet of 576 aircraft, and SECAF Talbott approved between 399 and 576 aircraft depending on available funding. SAC plans called for 11 B-52 wings assuming funding for 576 aircraft would ultimately be approved. SAC's B-52 basing plan followed the same strategy as its B-47 and B-36 wings with 45 bombers[1] and 15-20 tankers assigned to one overcrowded base. Many of these wings were planned for Northeastern bases and SAC again reorganized its NAFs on 13 June 1955 to better support bomber operations. Headquarters 8th AF moved from Carswell AFB, TX to Westover AFB, MA. The 2nd AF was now responsible for the southeast (including Texas), 8th AF was responsible for the northeast and central United States, while the 15th AF had the southwest and west.

By the late 1950s, SAC's B-52 basing strategy with 45 aircraft in each wing presented rich targets for more accurate and prevalent Soviet missiles. SAC planners were concerned that the Soviets could take out most of the bomber fleet with just a few missiles.

It was also impossible to launch so many B-52 aircraft from one runway within a survivable response time. Therefore, SAC began breaking up the large concentrations of aircraft in 1958 to complicate the Soviet targeting problem and increase available runways. Each B-52 wing was divided into three smaller wings of 15 bombers and 10-15 tankers.

In 1955, SAC began converting all B-36 wings and some B-47 wings to B-52 aircraft. (American Aviation Historical Society)

SAC also reorganized its B-47 and B-52 wings beginning in 1958 to support its new one-third ground alert program. The reorganization called for a new Deputy Commander for Operations (DCO) and Deputy Commander for Maintenance (DCM) to assist the Wing Commander to provide combat-ready crews and aircraft to support ground alert requirements. Maintenance was removed from the operational combat squadrons, and periodic maintenance squadron, and centralized into a newly formed Organizational Maintenance Squadron (OMS). Finally, the Air Base Group was redesignated as the Combat Support Group to emphasize its role in supporting the ground alert mission.

ARDC activated the 6592nd TS at Carswell AFB, TX on 1 March 1958 under the command of Colonel David M. Jones to support Category II R&D testing and coordinate all B-58 JTF activities. SAC activated the 3958th OETS at Carswell on the same day. This unit became an important member of B-58 JTF and was responsible for proper combat crew training development. SAC participation in the test program provided valuable experience toward the formation of B-58 training and operational units.

The first delivery of a YRB-58A (55-0661) to the 6592nd TS as part of the ARDC JTF occurred on 29 August 1958 at Carswell AFB, TX. It landed after a two-hour test flight on the Carswell runway, which was shared with the CVCFW factory where the aircraft was manufactured and initially tested. The aircraft was piloted by Air Force test pilot Major Leon Amos, who was the fourth Air Force pilot to be qualified in the B-58. Colonel Jones officially accepted the aircraft for the 6592nd TS. The 6592nd TS subsequently received 16 more YRB-58A (55-0660, 55-0663, 55-0666, 55-0670, 55-0671, 58-1007 to 58-1010, 58-1013, 58-1014, and 58-1016 to 58-1020) aircraft. They also received the seven initial B-58A (59-2428 to 59-2434) production aircraft to support R&D testing.

In 1959, SAC again realigned its NAF responsibilities. However, it was a relatively minor change moving one unit from the 8th AF to the 15th AF, but it gave the 8th AF responsibility for the Eastern United States, while the 2nd AF gained responsibility for the Central section, and the 15th AF retained the Western section. Along with this change, SAC realigned its ADs and gave them broader responsibilities. After the realignment, which included an increase in the headquarters staff from 17 to 25 personnel, each AD became responsible for three to five dispersed wings. The 3958th OETS at Carswell was redesignated as the 3958th OET Group (OTEG) on 1 September 1959 and assigned to the 19th AD at Carswell. The group had responsibility for training ground and air crews for the B-58.

Although the number of B-58 aircraft to be acquired remained in flux, SAC planners assumed that a three-wing B-58 fleet would be funded. They concluded that a small, fully operational B-58 force would drive the Soviet Union to develop some system of Mach 2.0 defenses that could cover all possible targets or accept the amount of destruction that a three-wing force could inflict. Most of the major targets west of the Urals would be vulnerable to an attack by refueled B-58s, with the B-52 and B-47 fleets providing mutual support for penetration of Soviet early warning and defense networks. With Mach 2.0 performance, the B-58 would greatly improve the overall strategic capability of SAC and require the Soviet Union to spend money at an unsustainable rate to defend against it. The first two wings would be based at Carswell AFB, TX and Bunker Hill AFB, IN, but SAC was not sure where the third one would be located. General Power, Lieutenant General McConnell, and other SAC officers argued that the third wing should be at Little Rock AFB, AR. Consequently, General Power asked his staff to start planning for the aircraft, along with a fleet of KC-135 tankers, to be located at Little Rock.[2]

In January 1960, the Air Force announced the first operational B-58 wing would be activated at Carswell AFB, TX. The 43rd BMW was subsequently transferred from Davis-Monthan AFB, AZ and established at Carswell on 15 March 1960. The move involved the transfer of only a few personnel from Davis-Monthan, but about 1,000 personnel assigned to the 3958th OETG at Carswell were reassigned to the 43rd BMW. Major General William H. Blanchard, SAC Director of Operations, read a message from General Power during the activation ceremony in which Power said the Air Force is "counting on the B-58... to contribute to a well-balanced aerospace force of the latest bombers and missiles. Activation of the 43rd BMW represents an important milestone in SAC's

B-58 aircraft of the 43rd BMW on the flightline at Carswell AFB, TX on 21 December 1962. (University of North Texas Libraries)

continuing endeavor to maintain an operational inventory of the most advanced weapon systems. I am confident... the 43rd Bomb Wing and other B-58 units to follow will help maintain the free world's most powerful deterrent to aggression and war."

On 1 August 1960, the 43rd BMW formally assumed B-58 operations responsibility and began Category III testing. The 6592nd TS was deactivated and eight of its 17 assigned YRB-58A aircraft (55-0661, 55-0666, 55-0671, 58-1013, 58-1014, and 58-1018 to 58-1020) were transferred to the 43rd BMW. Five of these aircraft (58-1013, 58-1014, and 58-1018 to 58-1020) were subsequently converted to B-58A aircraft and cycled back into the 43rd BMW. Two YRB-58A (55-0661 and 55-0671) aircraft were converted to TB-58A aircraft and assigned to the 305th BMW. YRB-58A (55-0666) was retained at CVCFW and the AFFTC for testing before being placed on display at Chanute AFB, IL.

Seven of the original YRB-58A aircraft (55-0660, 55-0663, 55-0670, 58-1007, 58-1009, 58-1010, and 58-1016) assigned to the 6592nd TS had already been returned to CVCFW. Three of these aircraft (58-1009, 58-1010, and 58-1016) were converted to B-58A aircraft and returned to the 43rd BMW. Two YRB-58A (55-0670 and 58-1007) aircraft were converted to TB-58A and assigned to the 43rd BMW, and one YRB-58A (55-0663) became a TB-58A assigned to the 305th BMW. YRB-58A (55-0660) was assigned to Kelly AFB, TX as a ground trainer before being placed on display. Two of the YRB-58A (58-1008 and 58-1017) aircraft crashed during testing.

The 43rd BMW also received the seven initial B-58A (59-2428 to 59-2434) production aircraft from the 6592nd TS and B-58A (59-2438) was assigned directly from the production line on 4 August. The first fully operational B-58A (59-2436) equipped with all tactical systems was assigned to the 43rd on 3 September 1960. Three more YRB-58A (58-1011, 58-1015, and 58-1021) aircraft previously retained at CVCFW for service tests were also subsequently converted to B-58A and delivered to the 43rd BMW. Two YRB-58A (55-0668, 55-0672) aircraft were converted to TB-58A. Another 26 B-58A aircraft were ultimately assigned to the 43rd BMW. Some of these deliveries were delayed due to unreliable BNS issues, maintenance difficulties, AGE shortages, and Category III test involvement. IOC was finally achieved in 1961 with the final aircraft delivered in May 1961.

The second wing to be equipped with the B-58A was 305th BMW at Bunker Hill AFB, IN. The 305th was initially planned as the first B-58 wing, but plans were changed to activate the 43rd BMW at Carswell first due to its proximity to CVCFW and the convenience that location provided for Category III testing. The 305th BMW had recently moved with its assigned B-47 and KC-97 aircraft to Bunker Hill in May 1959. A few months later its KC- 97 aircraft were converted to KC-135A. The wing began converting from B-47 when it received its first B-58A (59-2461) aircraft on 11 May 1961. The first TB-58A (55-0671) arrived on 30 June 1961. The wing was declared FOC in August 1962 and placed its first bomber on ground alert, along with the 43rd BMW, in September. It received the final production B-58A (61-2080) on 26 October 1962.

B-58A (59-2461) "Hoosier Hustler" arrives at Bunker Hill AFB, IN on 11 May 1961. (University of North Texas Libraries)

Each wing was organized for combat ready capability according to the structure implemented across SAC in 1958. The Wing Commander was the combat commander directly responsible for the operation and maintenance of all crews and aircraft. The DCO and DCM commanded a staff of crew members and maintenance personnel to meet daily training and Emergency War Order (EWO) requirements. The commander of a separate Combat Support Group was responsible for all base support activities such as transportation and personnel services.

Each wing required 70 certified combat-ready crews. Before a crew could be certified, it had to demonstrate unquestioned ability to perform a multitude of complex tasks associated with safely and reliably putting a nuclear weapon on target. Add to this the requirement of launching half a wing of multiengine jet bombers in less than 15 minutes.

Bringing crews to this state of readiness required considerable training in classrooms, simulators, and in the air. Crews were then required to maintain their certifications and combat-ready status through continuous ground and flight training.

Despite its original plans for three B-58 wings, SAC was forced to settle for only two wings as shown below due to funding issues. Each wing was officially designated as having 40 Unit Equipped (UE) aircraft. However, 50 aircraft were ultimately assigned to the 43rd BMW including 11 former test aircraft converted to B-58A and four to TB-58A. A total of six 43rd BMW aircraft were eventually destroyed. The 305th BMW had 53 aircraft including 49 new build B-58A and four TB-58A. A total of nine 305th BMW aircraft were eventually destroyed. Initially, the 43rd BMW location at Carswell proved very beneficial for Category III testing but the wing was also sharing space with 7th BMW B-52 and KC-135 aircraft. This eventually became unworkable and the 43rd BMW was moved in 1964 to Little Rock AFB, AR,[3] the location originally designated for SAC's third B-58 wing, replacing B-52E aircraft of the 4123rd Strategic Wing (SW) which were moved to Clinton-Sherman AFB, OK.

Unit	Components	Base	Aircraft	Notes
6592nd TS	N/A	Carswell AFB, TX	YRB-58A (Aug 58 – Jul 60) B-58A (Dec 59 – Jul 60)	Activated at Carswell AFB, TX on 1-Mar-58 to provide ARDC leadership support of the B-58 JTF. Deactivated when the 43rd BMW was activated on 15-Mar-60.
3958th OETS	N/A	Carswell AFB, TX (1958-60)	TF-102A (Mar-58-Aug-58) YRB-58A (Aug 58 – Jul 60) B-58A (Dec 59 – Jul 60)	Activated at Carswell AFB, TX on 1-Mar-58 to provide SAC support to the B-58 JTF. Shared B-58 aircraft with the 6592nd TS. Deactivated and absorbed when the 43rd BMW was activated on 15-Mar-60.
43rd BMW	63rd BS 64th BS 65th BS	Carswell AFB, TX (1960-64) Little Rock AFB, AR (1964-70)	YRB-58A (Aug 60 -Jul 62) B-58A (Aug 60 – Jan 70) TB-58A (Aug 60 -Jan 70)	Moved from Davis-Monthan AFB, AZ to Carswell AFB, TX and converted from B-47 to B-58 on 15-Mar-60. Moved to Little Rock AFB, AR on 1-Sep-64. Moved to Andersen AFB, Guam and redesignated 43rd SW when the B-58 was retired in Jan-70.
305th BMW	364th BS 365th BS 366th BS	Bunker Hill AFB, IN (1961-70) (Renamed Grissom AFB in 1968)	B-58A (May 61 – Jan 70) TB-58A (Jun 61 – Jan 70)	Converted to B-58 from B-47 at Bunker Hill in 1961. Redesignated 305th ARW when the B-58 was retired in Jan-70.

SAC ultimately assigned its B-58 force to two BMWs instead of the three initially planned.

On 8 December 1965, SECDEF Robert McNamara announced the planned phase-out of all B-58 and early B-52 (C, D, E, and F) aircraft by end of June 1970. This phase-out would also include base closures and unit relocations. While the B-58 aircraft at Bunker Hill (now called Grissom AFB) and Little Rock were scheduled for retirement, the bases would remain open, and the wings would take on new missions due to their illustrious histories dating back to WWII. The 43rd BMW was redesignated as the 43rd SW and replaced the 3960th SW at Andersen AFB, Guam. The 305th BMW became the 305th Aerial Refueling Wing (ARW) and remained at Grissom. The last B-58 aircraft were sent to the boneyard on 16 Jan 1970.

World Records

The B-58 was the world's first supersonic strategic bomber with a maximum speed above Mach 2.0. The B-58 set more records than any other combat aircraft and still holds the Tokyo to London speed record. It set 19 world records including 13 speed records, one endurance, three payload, and two altitude records with the aircraft flown in combat configuration by SAC combat crews. Five of the records were formerly held by the Soviet Union. The B-58 won all major international awards including the Bendix Trophy, Thompson Trophy, Blériot Trophy Cup, Mackay Trophy, and Harmon Trophy.

Untouchable

A B-58A (59-2442), named "Untouchable", from the 43rd BMW set six new Fédération Aéronautique Internationale (FAI) speed with payload records during Operation QUICK STEP I on 12 January 1961. The aircraft flew two laps of a 621 mile or 1,000 kilometer (which equaled 1,242 miles or 2,000 kilometers) closed loop course between Edwards AFB, CA and Yuma, AZ with payloads of 4,409.2 pounds (2,000 kilograms), 2,204.6 pounds (1,000 kilograms), and no payload (0 kilograms) with an average speed of 1,061.81 mph. The previous 0-kilogram record was 700.47 mph held by an F-101A. The previous 1,000- and 2,000-kilogram record was 639.18 mph set by a Soviet TU-104.

Untouchable crew was awarded the Distinguished Flying Cross. (U.S. Air Force)

B-58A (59-2442) "Untouchable" from the 43rd BMW set six new FAI speed with payload records during Operation QUICK STEP I on 12 January 1961. (U.S. Air Force)

The aircraft flew on a "tear-shaped" course requiring significant turns that resulted in it flying 1,328 miles total at an average speed of 1,300 mph. Top speed was 1,425 mph. The aircraft also set three 1000-kilometer records (0, 1,000, and 2,000 kilograms) by flying at an average speed of 1,200.19 mph. The previous record was 506.7 mph set by a Soviet TU-104 on 2 April 1960. The crew was awarded the Distinguished Flying Cross and included Maj Henry J. Deutschendorf (pilot),[1] Captain William L. Polhemus (navigator/bombardier), and Captain Raymond R. Wagener (DSO).

Roadrunner

B-58A (59-2441) "Roadrunner" improved on three of the records set three days earlier during Operation QUICK STEP I. (U.S. Air Force)

Another B-58A (59-2441), named "Roadrunner", from the 43rd BMW improved on three of these records two days later during Operation QUICK STEP II on 14 January 1961 with an average speed of 1,284.73 mph on the 1,000-kilometer course. It hit a top speed of 1,430 mph, burning off some of its paint in the process, and made the trip in 29 minutes.[2] The crew consisted of Major Harold E. Confer (pilot), Major Richard H. Weir (navigator/bombardier), and Captain Howard S. Bialas (DSO). Major Confer was awarded the Thompson Trophy on 28 February 1961, which was the first time in its 31-year history it was awarded for a record set by a medium bomber. According to General Thomas S. Power, CINCSAC, "The major significance of these record flights is that they have dramatically proved the capabilities of SAC's first operational supersonic bomber." Major Bialas became the first B-58 crewmember to log 1,000 hours in the B-58 in March 1964. He received a 1,000-hour pin from B.G. Reed, General Dynamics VP of Operations at the Fort Worth plant. Major Weir, then Lieutenant Colonel, became the first navigator/bombardier to log 1,000 hours in June 1965.

Firefly

A B-58A (59-2451), named "Firefly", set a new record for sustained speed at 2,000 kilometers per hour (kmph) for 30 minutes on 10 May 1961. The aircraft flew 669.4 miles (1,077 kilometers)[3] in 30 minutes and 43 seconds between 43,000 and 50,000 feet at an average speed of 1,302.07 mph (2,095 kmph). This bettered the record set by the QUICK STEP II crew on 14 January and nearly doubled the 700.047 mph over a 1,000-

kilometer closed loop record set by an RF-101 on 8 April 1959. The "Firefly" flew on a closed course starting and ending at Edwards AFB, CA. The course included corners at Needles, CA, Kingman, AZ, Morman Mesa, NV, and Lone Pine, CA giving three almost 90-degree turns. The aircraft slowed considerably while negotiating these turns

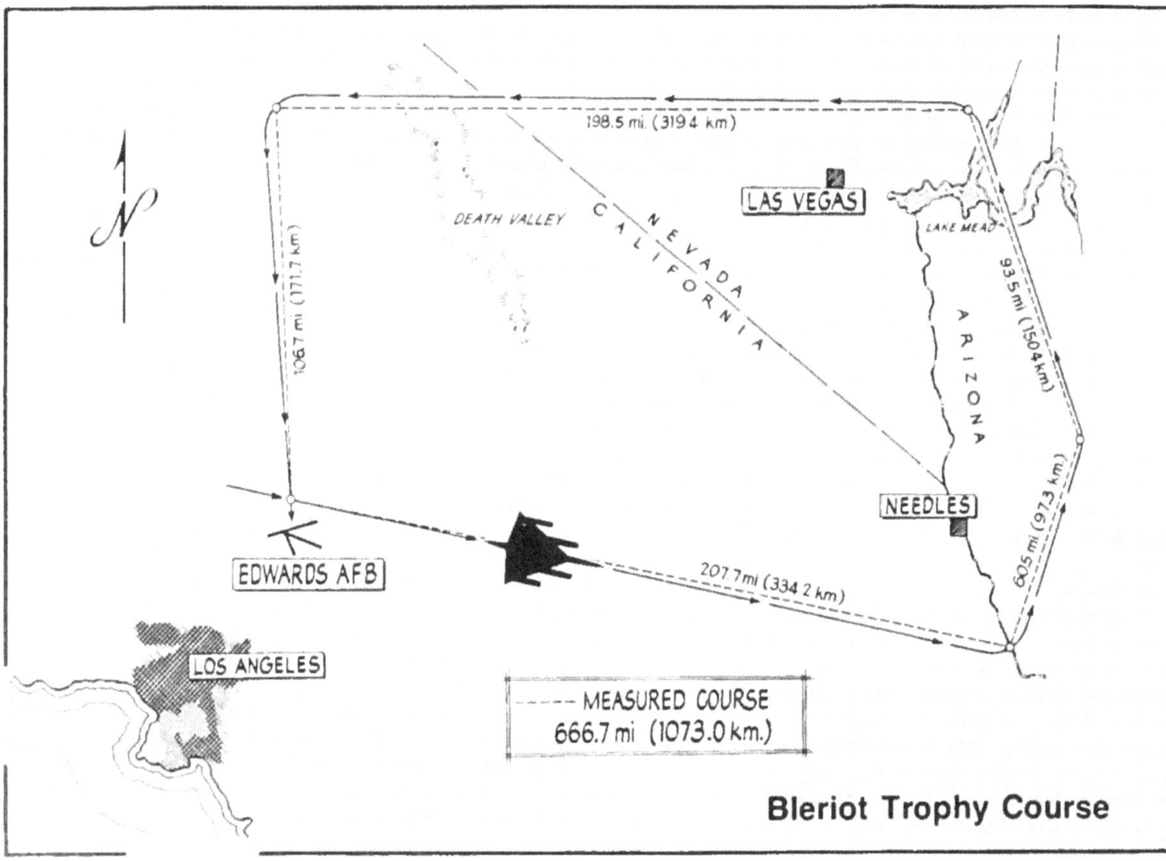

The pilot of the "Firefly" was awarded the Aero Club of France's Blériot Trophy Cup when it set a new record for sustained speed at 2,000 kilometers per hour (kmph) for 30 minutes on 10 May 1961. (U.S. Air Force)

making the actual average speed for the distance flown 1,365 mph. The crew included Major Elmer E. (Eugene) Murphy (pilot), Major Eugene Moses (navigator/bombardier), and First Lieutenant David F. Dickerson (DSO). Major Murphy was awarded the Aero Club of France's Blériot Trophy Cup which was established in 1930 by aviation pioneer Louis Charles Joseph Blériot to be awarded to an aviator who demonstrated flight at a speed of 2,000 kmph (1,242.74 mph) for 30 minutes.

The "Firefly" also set the FAI World Record for Speed Over a Recognized Course on 26 May 1961 on its way to the Paris Air Show for a 3,833.40-mile flight from Washington DC to Paris in 3 hours, 39 minutes, and 49 seconds at 1,048.68 mph average. The aircraft also completed a 3,625-mile flight from NYC to Paris in 3 hours, 14 minutes,

B-58A (59-2451) "Firefly" crew was awarded the Aero Club of France's Blériot Trophy Cup for demonstrated flight at a speed of 2,000 kmph (1,242.74 mph) for 30 minutes. Unfortunately, this crew was lost in a crash just three weeks later. (U.S. Air Force)

and 44.53 seconds at 1,089.36 mph average, which is about one-tenth of Charles Lindbergh's 33 hour and 32-minute flight in the "Spirit of St. Louis" on 20 May 1927. The flight ended in Paris on the exact spot where Lindbergh had cut the engine on his pioneering flight. The crew comprised of Major William R. Payne (pilot), Captain William L. Polhemus (navigator/bombardier) and Captain Raymond R. Wagener (DSO) won the seventh Mackay Trophy for this flight. Payne, then a Lieutenant Colonel, also received the 1962 Harmon Trophy on behalf of the crew for outstanding contribution to aviation from President Kennedy during a White House ceremony on 18 October 1962. He was

Blériot Trophy Cup. (University of North Texas Libraries)

Convair B-58A (59-2451) "Firefly" on the taxiway at the Paris Air Show. (U.S. Air Force)

B-58A (59-2451) "Firefly" at the Paris Air Show. Tragically, this aircraft and its crew were lost on 3 June 1961 during a demonstration at the Paris Air Show. (U.S. Air Force)

also the first to become combat ready and later became the first pilot to log 1,000 hours in the B-58 while flying a TB-58A on 17 July 1964. He received a 1,000-hour pin from

B.G. Reed, General Dynamics VP of Operations, at the Fort Worth plant upon completing the flight.

B-58A (59-2451) "Firefly" lands at Paris Air Show after its 26 May 1961 record flight. (University of North Texas Libraries)

Unfortunately, the "Firefly" disintegrated in flight during the Paris Air Show at Le Bourget Airport, France on 3 June 1961.[4] The crew, comprised of Major Elmer E. Murphy (pilot), Major Eugene Moses (navigator/bombardier), and First Lieutenant David F. Dickerson (DSO), was the same crew that earned the Blériot Trophy Cup just three weeks earlier. They had returned to Paris to receive the trophy in a ceremony presented by Blériot's widow. They were tapped to perform a flight demonstration following the ceremony while the primary crew appeared on a local television show. The bomber entered a cloud bank while performing a slow roll. The pilot lost his visual reference, and the roll caused the attitude indicator to exceed its limits. Disoriented and without instrument flight capability, the B-58 crashed. An eyewitness said the plane appeared to explode in flight after making a "barrel roll."

Cowtown Hustler

Operation HEAT RISE crew before flight. (National Museum of the U.S. Air Force)

A B-58A (59-2458), named "Cowtown Hustler", (call sign "Tall Man Five-Five") from the 43rd BMW broke three speed records during Operation HEAT RISE on 5 March 1962 for a round-trip flight of nearly 5,000 miles from Los Angeles to New York City and return. The total flight time was 4 hours, 41 minutes, and 14.98 seconds at an average speed of 1,044.97 mph with three in-flight refuelings required. The total time included a 25-minute aerial refueling over the Atlantic Ocean before the return flight to Los Angeles. Top speed was logged at 1,450 mph and the aircraft skin reached 225 degrees. This topped, by over two hours, the previous record set in 1957 by an RF-101C fighter of 6 hours and 46 minutes at an average speed of 721-plus mph. The Los Angeles to New York City (outbound flight) flight was completed in 2 hours and 58.71 seconds at 1,214.65 mph average. This beat the previous best time of 2 hours and 48 minutes at an average speed of 871-plus mph set by an F4H-1 fighter bomber.

The New York City to Los Angeles (return flight) was completed in 2 hours, 15 minutes, and 50.80 seconds at 1,081.81 mph average speed. This was the first transcontinental flight in history at a speed faster than the rotational speed of the earth, beating the Sun's westward track across the country. It topped the previous best time of 3 hours and 36 minutes at an average of 677-plus mph set by an RF-101C fighter. The crew, comprised of Captain Robert G. Sowers (pilot), Captain Robert MacDonald (navigator/bombardier), and Captain John T. Walton (DSO), earned the Mackay Trophy for the flight overall and the Bendix Trophy for the outbound flight. They were the first bomber crew to ever win the Bendix Trophy. They also received the Distinguished Flying Cross awarded by General Power. SAC typically flew the B-58 supersonic missions over isolated areas to prevent damage from the inevitable sonic boom, but this time they flew the same Los Angeles to New York City route as Jimmy Doolittle who had previously won the Bendix Trophy for making the trip at 223 mph. This took the aircraft over heavily populated areas. Several claims for broken windows were filed along the route.

Operation HEAT RISE crew posing in front of another B-58A (59-2447). (National Museum of the U.S. Air Force)

The second B-58A (59-2459), call sign "Tall Man Five-Six", involved in Operation HEAT RISE developed mechanical problems after passing New York City for the return flight and was forced to return to Carswell. It crashed later that day on take-off from Carswell and according to one report, "The speedy bomber veered to the right on the takeoff run, slammed through a 10-foot chain link fence and disintegrated atop an ammunition bunker, about 75 yards short of Lake Worth." Another report said, "The aircraft lost one wing and went into a 6,000-

The Operation HEAT RISE crew won the Bendix Trophy (shown), Mackay Trophy, and Distinguished Flying Cross. (University of North Texas Libraries)

foot skid that left the aircraft upside down and burning on a partially buried ammunition dump. One crewmember ejected but was not thrown clear of the wreck and fire." Its crew, Captain Robert Eugene Harter (pilot), Captain Jack De Voll Jones (navigator/bombardier), and First Lieutenant James Thomas McKenzie (DSO) were all killed. This was the eleventh B-58 loss.

Harmon Trophy

Another B-58A (59-2456) acquired two more FAI altitude records on 14 September 1962 during a zoom climb over Edwards AFB, CA carrying a payload of 5,000 kilograms to 85,360.66 feet and won the 1962 Harmon Trophy. It set the FAI World Record for Altitude in both the 2,000-kilogram (4,409.25 pounds) and 5,000-kilogram (11,023.11 pounds) classes breaking records previously held by the Soviets. The Soviet records were 50,252.2 feet with a 5,000-kilogram payload set on 16 September 1959 and 66,170.72 feet with a 2,000-kilogram payload set on 29 July 1959. The Air Force

Major Fulton in the cockpit of B-58A (59-2456) Harmon Trophy winner. (U.S. Air Force)

Systems Command flight crew included Major Fitzhugh L. (Fitz) Fulton, Jr. (pilot) who was also the first Air Force flight test pilot to fly the B-58 in early 1957, Captain William R. Payne (navigator/bombardier) and Charles R. Haines (civilian flight test engineer).

Greased Lightning

Perhaps the most famous record-setter was the B-58A (61-2059), named "Greased Lightning", from the 305th BMW which set five records on 16 October 1963 during Operation GREASED LIGHTNING. The aircraft set, and still holds, the world speed record flying 8,028 miles from Tokyo to London in 8 hours, 35 minutes, and 20.4 seconds at 938 mph average. This cut the previous record of 17 hours and 42 minutes held by a British jet by more than half. The flight demonstrated SAC's ability to divert a B-58 to any spot on the globe and do it supersonically, according to SAC headquarters. The mission served as an operations and training objective involving many units, crews, and support personnel. Three Hustlers, their crews, and ground support personnel were deployed to Okinawa. Several RB-47 weather scout aircraft patrolled the flight path to provide real-time weather updates. The flight required five in-flight refuelings which required aircraft to slow down to subsonic speed and affected the average speed. KC-135 tankers departed from eight different bases in six different states along the route to support the flight. All but one of the five refuelings was executed uneventfully at 26,000 feet. However, a storm over anchorage forced one B-58 to execute a high-speed

B-58A (61-2059) "Greased Lightning" after its Tokyo to London record flight. (University of North Texas Libraries)

refueling at 28,000 feet, which was a difficult maneuver and required almost eight minutes due to severe turbulence.

The routes flown involved two routine runs: GLASS BRICK which was a regular practice mission to the Far East; and ALARM BELL which was a similar mission to Spain. SAC planned a switch in destination and flight path as a curve ball to keep the crews on their feet. Three aircraft took off on the GLASS BRICK run from Kadena AB, Okinawa and flew over Japan, to the Aleutian Islands in Alaska. From that point one aircraft (no one knew which one beforehand) was diverted to London. One aircraft landed at Eielson AFB, AK and the other was routed to Chicago and then back home to Bunker Hill.

The remaining aircraft (61-2059) flew over Northern Canada, Greenland, Iceland, and London before landing at RAF Greenham Common in England. The Tokyo to Anchorage leg was completed in 3 hours, 9 minutes, and 42 seconds at 1,093.44 mph average. The Anchorage to London leg was completed in 5 hours, 24 minutes, and 54 seconds at 826.91 mph average making it the longest supersonic flight in history at that time. The crew included Major Sidney J. Kubesch (pilot), Major John Barrett (navigator/bombardier), and Captain Gerard Williamson (DSO).

Additional records would be set by the B-58 before its operational career ended, but these would not be officially logged with the FAI and thus not placed on the record books. One B-58, for instance, which returned to Bunker Hill AFB from routine deployment to Okinawa, set two world speed records in the process. This aircraft covered the

B-58A (59-2458) Bendix Trophy winner arrives in Los Angeles on 5 March 1962. (University of North Texas Libraries)

B-58A (59-2458) Bendix Trophy winner sits on exhibition during the Mackay Trophy Presentation at Carswell on 8 August 1963. (University of North Texas Libraries)

B-58A (59-2458) Bendix Trophy winner arrival ceremony in Los Angeles on 5 March 1962. (University of North Texas Libraries)

General Power greets the Bendix Trophy crew in Los Angeles on 5 May 1962. (University of North Texas Libraries)

2,858 miles between Anchorage and Chicago in 5 hours, 23 minutes, and 20 seconds at an average speed of 530.88 mph. The overall Tokyo-to-Chicago distance of 6,371 miles was flown in 8 hours, 36 minutes, and 38 seconds averaging 740.76 mph. At the time, no official records existed for flights between Tokyo and Chicago or between Anchorage and Chicago. Lieutenant Colonel George Andrews, Major Joseph Guastella, and Captain Clifford Youngblood crewed the aircraft on this little-known record breaker.[5]

Operations

Cold War Operations

The B-58 quickly became a very effective aircraft. Approximately 103 aircraft were fielded and assigned to two bomb wings. By the end of 1963, Air force crews had made over 10,500 flights in the B-58, amounting to about 53,000 flying hours. Of these hours, 1,150 were supersonic and 375 were at Mach 2.0. By the mid-1960s, with the B-47 nearly phased out, the B-58 and B-52 were the only two bombers programmed into the 1970s. They formed the all-important mix of aircraft and missiles which Air Force leaders unanimously agreed was a critical factor in maintaining a credible deterrent force. As the newer of the two and better suited for low-level penetration techniques, the B-58 appeared then to be more important than its few numbers would indicate. It also seemed to resemble, to some degree, the next generation of manned bombers while the Air Force pursued development of the Advanced Manned Strategic Aircraft (AMSA), which would eventually evolve into the B-1 bomber.[1]

Wing Operations. The B-58A bomber force was equally divided between the 43rd BMW and the 305th BMW with each wing having at least 40 aircraft. During normal operations 50 percent of the assigned aircraft and crews were on ground alert. The remaining aircraft and crews were left to fly SAC exercises and training missions. Each available aircraft typically flew every other day on 6- to 8-hour training missions. Each wing also had four TB-58A aircraft assigned for pilot instrument training. Each of these aircraft typically flew three training sorties in each 24-hour period.[2]

SAC training missions approximated as closely as possible the actual profile of an EWO mission, while satisfying combat crew and training requirements. Each training mission typically contained navigation legs, a refueling, and one or more radar-scored simulated bomb drops from either high or low altitude. Training missions could also include a supersonic leg. Thousands of simulated bomb runs against fixed and mobile radar bomb scoring units proved the very small

Typical B-58 Training Mission Profile

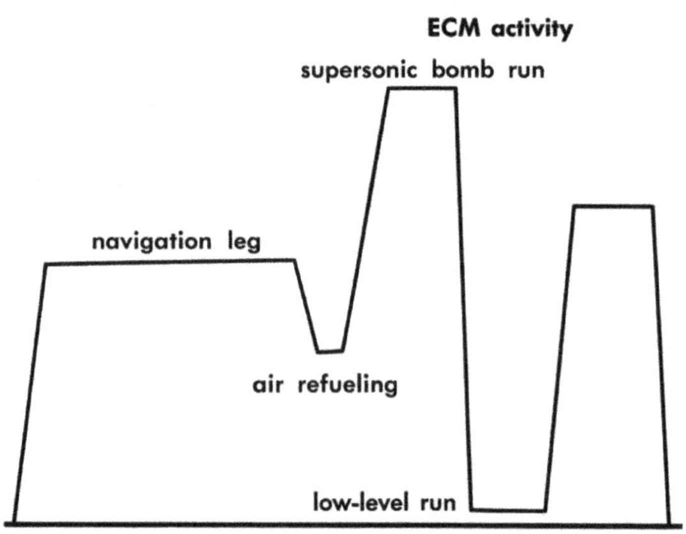

All SAC B-58 crews could fly this profile effectively against Soviet, satellite, or Chinese targets. (U.S. Air Force)

Circular Error Probable (CEP) of the B-58 at both high and low altitudes. These missions were flown in all weather conditions, day or night. All SAC B-58 crews could fly this profile effectively against Soviet, satellite, or Chinese targets. EWO missions differed in that they used buddy refueling and an overseas recovery base.[3]

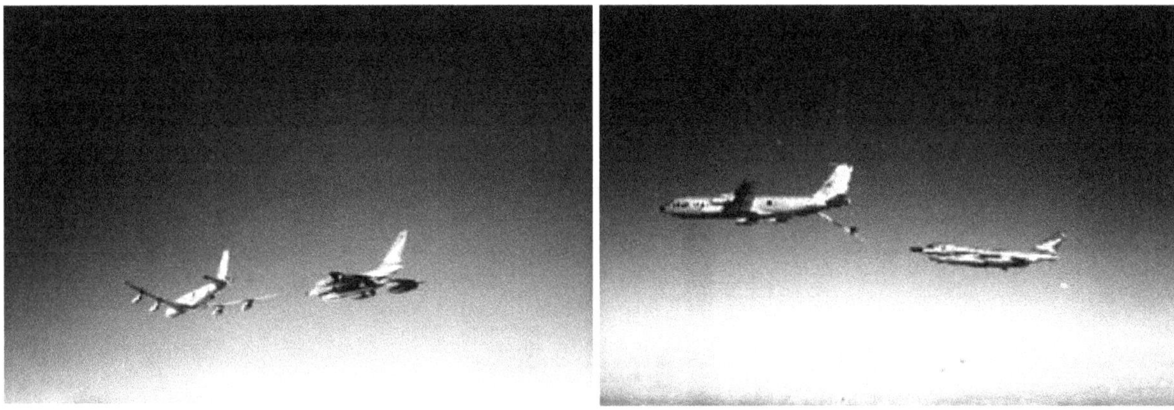

A B-58A approaches a KC-135 tanker during refueling operations. (University of North Texas Libraries)

Wing combat readiness was monitored constantly. All combat crews were subject to regular standardization and evaluation checks by the SAC Combat Evaluation Group. Each wing was also tested by annual no-notice Operational Readiness Inspections (ORI) in which the entire wing was exercised to fly a simulated initiation of hostilities. Added to these evaluations were BAR NONE exercises in which the wings were required to launch all available aircraft against unfamiliar targets on a strict take-off schedule. SAC also exercised the B-58 periodically with the North American Air Defense Command (NORAD).[4]

SAC also deployed aircraft from each B-58 wing to either England and Spain or Guam and Okinawa. The two wings alternated between these forward-based training missions. These deployments allowed force dispersal and gave a limited forward B-58 capability. They also gave base support personnel training in aircraft and weapons handling.

SIOP. The B-58 fleet was an integral part of the SIOP. They stood alert loaded with nuclear weapons and ready for combat. As of September 1962, SAC had a requirement to have nuclear weapons loaded on alert B-58 bombers. These aircraft were kept on constant alert ready status in support of the SIOP. Each B-58 could carry up to four thermonuclear weapons on pylons and one thermonuclear weapon in the TC pod.

The B-58, a medium bomber, was really a 'heavy lift' machine–capable of carrying its own weight plus twice its weight in fuel and weapons. The lower "fuel pod" was filled with fuel. The upper "bomb pod" fit inside the top opening of the fuel pod. Four bombs were also mounted on pylons attached to hardpoints under the wing. On a SIOP mission, the B-58 first burned fuel from the fuel pod. It then jettisoned that pod leaving the thermonuclear bomb pod still attached for subsequent bomb release on target. The bomb pod consisted of three components. The front and aft sections contained fuel while the middle section contained the thermonuclear bomb.[5]

Ground Alert Program. By 1957, the Soviet Union was developing much improved intercontinental ballistic missiles and SAC planners began exploring concepts for getting its force airborne within 15 minutes after detecting a missile attack. They settled on a ground alert program in which at least one-third of the bomber and tanker force would be loaded with crews standing by for immediate take-off. This concept was initially applied to B-47 and KC-97 aircraft and eventually expanded to the B-52 and KC-135 force.

Convair B-58A (59-2435) with a TC fuel and weapons pod and four nuclear weapons on pylons. (U.S. Air Force)

SAC conducted three test programs at B-47 units to verify the operational viability of the program and to identify improvements. The first, Operation TRY OUT, was conducted from November 1956 through March 1957 at Hunter AFB, GA. It proved the concept viable and identified numerous changes needed. SAC conducted two additional tests to perfect these changes. Operation WATCH TOWER was conducted at Little Rock AFB, AR from April to November 1957 and Operation FRESH APPROACH was conducted at Mountain Home AFB, ID in September. These tests all proved very successful and, although there were still some details to work out, General Thomas Power,

CINCSAC, directed that ground alert operations would start at several CONUS and OCONUS bases on 1 October 1957.

General Power penned a memorandum to all SAC alert crew members on 9 November 1957 to mark the occasion. He wrote, "As a member of SAC's Alert Force you are contributing to an operation which is of the utmost importance to the security and welfare of this nation and its allies in the free world... the only way of insuring the survival of some of SAC's combat capability, even in the case of the most unexpected and massive attack, is our Alert Force. As long as the Soviets know that, no matter what means they may employ to stop it, a sizable percentage of SAC's strike force will be in the air for

Two crewmembers quickly board a YRB-58A at night during testing on 16 September 1959. (University of North Texas Libraries)

the counterattack within minutes after they have initiated aggression, they will think twice before undertaking such aggression. For this reason, it is my considered opinion that a combat-ready Alert Force of adequate size is the very backbone of our deterrent posture."

SAC achieved its one-third ground alert goal in May 1960. Alert bombers (B-47 and B-52) and tankers (KC-97 and KC-135) were loaded and ready to take off upon detection of incoming Soviet missiles. SAC forces were on alert at 46 bases across the United States and 20 OCONUS bases. This greatly increased the number of runways available to support alert take-offs. SAC's reaction capability was further enhanced through a new Minimum Interval Take-Off (MITO) procedure implemented beginning in mid-1960. This capability was tested in the spring of 1960 at several bases under Project OPEN ROAD and allowed the alert force to become airborne within 15 minutes giving the U.S. the ability to conduct massive retaliatory attacks on the Soviet Union.

SAC began operating a new underground command post at SAC headquarters along with EC-135 Airborne Command Post aircraft, called LOOKING GLASS, to provide around the clock control of all SAC nuclear weapons. SAC also implemented a new High Frequency (HF) communication system, called SHORT ORDER, that allowed positive control of the alert bombers enroute to the target. Using positive control procedures, the bombers flew to a designated point outside enemy territory and awaited a Go-Code before proceeding to their targets. If no code was received, they returned to home station.

On 28 March 1961, President Kennedy requested funds to strengthen and protect the strategic deterrent force and strengthen the ability to wage limited war. He ordered 50 percent of SAC's B-52 and B-47 bombers, along with KC-97 and KC-135 tankers, to be placed on ground alert. He also directed accelerated B-47 phase out to provide the crews for expanded ground alert. SAC achieved the 50 percent target in July with up to 519 B-47 and B-52 bombers on alert. This number peaked at 625 B-47, B-52, and B-58 bombers on alert by the end of 1962.

On 6 September 1961, SAC officially stated its desire to establish a B-58 reflex capability in an overseas area during the 1963-1970 period. SAC requested Air Force approval of the operational concept and requirements. Approval was granted on 27 September, and AFLC was then asked to appraise the support aspects and to begin the preliminary planning action. B-58 reflex capability was established at initial overseas areas by 1 July 1963 and expanded to several other areas thereafter. The two wings alternated participation in these forward-based training missions. Such deployments dispersed the B-58 force, gave SAC a limited forward B-58 capability, and gave base support personnel training in aircraft and weapons handling.[6]

The B-58 joined the alert force when aircraft of the 43rd BMW and the 305th BMW began assuming ground alert in September 1962. The B-58 was expected to be among the first ground alert aircraft over hostile targets due to its ability to take off quickly and its supersonic dash capability. Like the B-52 bombers, the B-58 aircraft were eventually modified with quick-start cartridges and were fully loaded ready for MITO launches within 15 minutes. Regular evaluation demonstrated that the B-58 force could be airborne well within the period expected from the Ballistic Missile Early Warning System (BMEWS).

An alert crew runs to a B-58A (59-2430) which has been "cocked-on" alert. (American Aviation Historical Society)

SAC kept 50 percent of all combat-ready B-58 aircraft on alert. While on alert they could not receive scheduled maintenance or fly proficiency or training sorties. Crews typically pulled alert duty for seven days every three weeks. They were housed in a secure alert facility and ready to go at a moment's notice. In effect, the alert force was detached from the parent wing and assigned to SAC headquarters.

SAC began reducing its alert force in 1963 with the phase-out of B-47 bombers directed by President Kennedy. In 1964, SAC had a total of 464 B-47, B-52, and B-58 bombers on alert. It also replaced its B-47 ground alert force at Andersen AFB, Guam with B-52 aircraft that same year. B-47 ground alert across the command terminated on 11 February 1966 and only 301 B-52 and B-58 aircraft remained on alert. By 1967, SAC had reduced its alert force to 40 percent with 223 B-52 and B-58 bombers on alert due to Vietnam War commitments.

SAC implemented a new B-52 dispersal program in 1968 based on its earlier success with the B-47 force. Aircraft were dispersed during times of crises to various military and civilian airfields and placed on ground alert to spread out the force and complicate the enemy's targeting problem. On 20 February 1969, SAC began testing a new B-52 satellite basing program at Homestead AFB, FL (TAC base) using aircraft relocated from Ramey AFB, Puerto Rico. A small maintenance and support contingent deployed with the aircraft, while the crews rotated back and forth between the satellite base and their home base. The test was completed by 20 May, and several wings were brought into the program on 1 July. These programs were similar except the aircraft were regularly deployed to satellite bases whereas the dispersal program was used only during times of crises. Both programs multiplied the available runways and increased the probability of launching the entire alert force before Soviet missiles could strike. B-58 units did not participate in these programs since the aircraft could take-off much faster than the B-52 and all of the alert aircraft at each of the two main bases could easily become airborne within 15 minutes.

Low-Level Operations. The B-58 was well-suited for low-level operations although it was designed as a high-altitude supersonic bomber. The small size of the B-58 was a huge advantage for the low-level mission. Its wingspan was just one third of the B-52 and its length was about one half. This resulted in a radar reflectivity of about one thirteenth of the B-52. The high aspect ratio, delta design, wing loading, control system, and arrangement of the B-58 resulted in a superior ride over the B-52 at low-level penetration altitudes.[7]

A B-58A flies a high-speed low-level pass over Edwards AFB, CA in May 1963. (American Aviation Historical Society)

The B-58 was much less susceptible to low-level turbulence than the B-52. The strong delta wing and short profile of the Hustler made it relatively free from the oscillation effects of low-level turbulence experienced by the B-52. Like the B-47 and the B-52, the B-58 started showing signs of structural fatigue after numerous hours of low-level flying. However, in 1967 all B-58 aircraft underwent extensive inspection and repair of all primary and secondary structures to mitigate the effects of low-level turbulence.[8] With its high speed, low radar signature, and ability to withstand heavy turbulence at low altitude, the B-58A was very effective at low-level penetration. Although the planned

General Dynamics AN/APN-170 TFR installation was tested in the B-58 but later cancelled due to funding, a new highly accurate Radar Altimeter was installed in 1969 allowing the B-58 to fly even lower.[9]

Convair YRB-58A (55-0660) during low level flight. (U.S. Air Force)

To prove its low-level capability YRB-58A (58-1015), piloted by Convair test pilot B.A. Erickson, flew three low-level penetration flights from Carswell AFB, TX to Edwards AFB, CA on 18 September 1959. The aircraft flew at an average speed of 703 mph (nearly speed of sound) in 1 hour and 46 minutes elapsed time never more than 500 feet above ground. It continued westward to the Pacific shoreline, completing the total run of 1,400 miles in two hours flat. A portion of the flight over New Mexico was conducted at below 200 feet. The aircraft essentially disappeared from the tracking radar. This superb low-level performance easily allowed the B-58 to sneak under an enemy radar.

B-58 test engineers mounted a space-age "cowbell" on the aircraft so they could find it during simulated bomb runs over the Kirtland AFB range. Radar operators found it extremely difficult to see the Hustler on their ground station screens so Convair was forced to install a radar beacon to make detection easier. Even though the radar operators knew the second the B-58 would make its run, the course, and altitude, they found it

YRB-58A (58-1015) flew three low-level penetration flights from Carswell AFB, TX to Edwards AFB, CA on 18 September 1959 to prove the aircraft's low-level capability. (American Aviation Historical Society)

virtually impossible to spot and track the aircraft. This was caused by the altitude (over 60,000 to less than 500 feet) and speed (subsonic to over Mach 2) at which the B-58 could bomb accurately. Once the beacon was added and the radar operators could track the bomb run, they discovered the Hustler was scoring an impressive number of simulated drops right on target using its highly accurate BNS.

Several factors contributed to the B-58 bomber's low radar reflectivity. The aircraft's smooth skin due to bonding as opposed to riveted surfaces; a pencil thin body due to tandem crew seating as opposed to side-by-side seating; and its area rule 'coke bottle' shape, all tended to mask reflective surfaces from enemy radars. The engine variable inlet spikes also contributed in a big way. Rotating jet engine blades provide a very high reflective surface for enemy search and tracking radars, especially when inbound to the target. However, the spikes on the B-58 shielded the engine blades from enemy radar. No wonder that Nike site radar operators quite often asked an inbound Hustler to 'show itself' on a bomb run, so it could be picked up on radar. To 'show itself' meant that the pilot had to climb to an altitude where the Hustler showed up as a small but definite blip on their radar screens and was no longer obscured by radar ground clutter returns. The B-58 was the only aircraft that Nike operators had to make this request to 'show itself.'[10]

Competitions and Exercises

SAC Bombing Competition. In 1948, General George Kenney, SAC Commanding General, decided to hold the first SAC Bombing Competition from 20-27 June at Castle AFB, CA hoping to stimulate interest in improving bombing accuracy. The competition was so successful at increasing competitive spirit among the crews and bombing accuracy, that General LeMay made it an annual event when he took command of SAC in 1949. It later evolved to include navigation and was renamed the SAC Bomb-Nav Competition in 1951, although it was still typically called the SAC Bomb Comp.

Two B-58A Hustlers from the 43rd BMW entered the competition held from 13-15 September 1960 for the first time just six weeks after the B-58 fleet became operational. The aircraft flew from Carswell to Bergstrom AFB, TX, where the competition aircraft

Two B-58A Hustlers from the 43rd BMW entered the SAC Bomb Comp held from 13-15 September 1960 for the first time just six weeks after the B-58 fleet became operational – they performed admirably. (American Aviation Historical Society)

were being staged. The B-58 crews competed against six B-47 and six B-52 wings and SAC's most experienced and capable combat crews. The primary B-58A (59-2429) was crewed by Major Harold E. Confer (pilot), Major Richard H. Weir (navigator/bombardier), and Captain Howard S. Bialas (DSO). The secondary crew was Major Henry Deutschendorf (pilot), Captain William L. Polhemus (navigator/bombardier), and Captain Ray R. Wagener (DSO). The secondary crew arrived in the back-up B-58A (59-2430) but ultimately both crews flew only the primary aircraft during competition.

Although the 11th BMW, a B-52 wing from Altus AFB, OK, won the Fairchild trophy for best combined score in alert activities, bombing, navigation, ECM, and aerial refueling the B-58 performed admirably. The B-58 crews placed 5th overall and totaled 1,046 points, which was just 137 points behind the first place B-52. They completed all missions including night refueling, high and low-level bomb runs, and alert response. Major Confer's B-58 won the radar bombing trophy with a pair of first place radar bombing sorties. It also scored 100 of 100 points for the quickest response from ground alert – 2 minutes 10 seconds – almost half the time for most other aircraft in the competition. Major Deutschendorf's crew bettered this time the following day, clocking 2 minutes and 5 seconds, again earning 100 of 100 points. His crew also scored 100 of 100 points in ECM and refueling.

Major General Keith K. Compton, SAC Deputy Commander of Operations, said in his speech at the awards ceremony on 16 September, "... in my opinion the 1960 Combat Competition marks a definite turning point in the history of SAC's growth and progress. This year's competition was a definite departure from past competitions, in many phases, and it certainly indicates that the manned force in SAC has come into a new era. The superb bombing of the B-58 force in this competition definitely proves that this is now even more apparent when you realize that SAC accepted the B-58 weapon system only six short weeks ago. This competition has proved beyond doubt that this weapon system is a potent alert vehicle with positive, accurate, striking power..."

The two B-58 wings were too involved in training in 1961 to compete in the Bomb Comp. No Bomb Comp was held in 1962 due to Cuban Missile Crisis operations. The competition was also cancelled in 1963 due to increased operational commitments and cost reductions. However, General Power decided to award the Fairchild Trophy to the wing with the best overall combat capability record for Fiscal Year 1963. A special board of officers reviewed the records of all wings and selected the 2nd BMW, a B-52 unit from Barksdale AFB, LA as the Fairchild Trophy winner. The competition was again cancelled in 1964 due to cost reductions. This time the Fairchild Trophy was awarded to the 70th BMW, a B-52 unit from Clinton-Sherman AFB, OK, using the same process as in 1963.

The Bomb Comp continued in 1965 and was held from 12-18 September. One crew participated from each of the 44 bomb wings (two B-58, five B-47, and 37 B-52). The

454th BMW, a B-52 unit from Columbus AFB, MS won the Fairchild Trophy. This was the last competition for the B-47. A B-58A from the 305th BMW placed first in navigation and third overall. The crew consisted of Major Mackie L. Sorrell (pilot), Captain Donald A. Itzen (navigator/bombardier), and Major George R. Green, Jr. (DSO). Another B-58A from the 43rd BMW placed in the upper 25 percent in the overall competition. The crew was Major Prentice H. Farris (pilot), Captain William G. Potts (navigator/bombardier), and Captain William R. Wirth (DSO).

A B-58A from the 305th BMW placed first in navigation and third overall in the 1965 competition. (American Aviation Historical Society)

By the time the fifteenth competition was held from 2-8 October 1966, the meet was becoming a predominantly B-52 affair. Participants included one crew each from 35 B-52 wings and two B-58 wings. Three Royal Air Force (RAF) Vulcan crews from the UK also participated. With the addition of international competition, the event was hereafter often referred to as the World Series of Bombing. The 8th AF was the big winner with the first four places in the overall competition and its 19th BMW, a B-52H unit from Homestead AFB, FL, won the Fairchild Trophy.

The Bomb Comp was again cancelled in 1967 due to operational commitments, and no Fairchild Trophy was awarded. No competition was held in 1968 due to Southeast Asia (SEA) combat operations. It continued the following year from 5-15 October 1969 and included 22 B-52, two B-58 wings, and three RAF crews flying Vulcan bombers. Several

B-52D wings did not participate due to SEA combat operations. This was also the last competition for the B-58. The 319th BMW, a B-52H unit from Grand Forks, ND, won the Fairchild Trophy.

Broken Arrows

The term Broken Arrow was used within the Defense Department, and consequently SAC, to describe an incident involving a nuclear warhead. Typically, it involved aircraft crashes or accidents with a nuclear weapon onboard. While SAC reported several broken arrows over the years on other aircraft, only one was reported involving a B-58 aircraft. This broken arrow occurred on 8 December 1964 at Bunker Hill AFB, IN.

B-58A (60-1116). This aircraft was completely destroyed by fire when the landing gear collapsed after the aircraft slid off the icy runway during an exercise. Several B-58A and KC-135A aircraft were taxiing during a no-notice ORI alert exercise. The ORI had been called during an extreme winter storm. The alert pad, taxiways, and runways were all covered with ice and snow and visibility was severely limited due to blowing and falling snow.

B-58A (60-1116), fully loaded with five nuclear weapons, was the first to leave the alert pad. The pilot was eager to be the first to the runway. However, he quickly got lost in the blowing snow and taxied past the runway. The tower soon alerted him, and the pilot turned around and headed back. By the time he reached the runway he was last in line and positioned directly behind a KC-135A tanker. The combination of jet blast from the tanker, the icy surface conditions, and the power applied to the B-58A while attempting to turn onto the runway, caused the pilot to lose control and the aircraft slid off the left-hand side of the taxiway. The left main landing gear passed over a flush-mounted taxiway light fixture and 10 feet further along in its travel, grazed the left edge of a concrete light base. Ten feet further, the left main landing gear struck a concrete electrical manhole box, and the aircraft caught on fire.

When the aircraft came to rest, all three crew members onboard raised their canopies and began abandoning the aircraft. Captain Larry Johnson (pilot) and Captain Roger Hall (DSO) egressed with only minor injuries. Johnson was able to crawl down the nose of the aircraft with flames licking at him and jump to the ground. Hall climbed over the right wing through the fire that was now raging around the aircraft and jumped to the ground. His flight suit caught fire and was extinguished after rescuers rolled him in the snow. He received some minor burns along with a dislocated shoulder. Captain Manuel "Rocky" Cervantez (navigator/bombardier) had nowhere to go. The aircraft was now engulfed in flames, and he had no way to crawl out of his compartment. With no other option he ejected his escape capsule, which shot up about 100 feet and impacted like a rock 548 feet from the aircraft.[11] He did not survive.

The five nuclear weapons on board were engulfed in flames and resulted in radioactive contamination of the crash area. Some portions of the weapons were melted and burned but did not detonate. One weapon was extinguished by placing it in a shallow trench and covering it with sand. All radioactive contamination was removed and buried with the aircraft on a remote location of the base. The destroyed nuclear bombs were taken to disposal facilities.

Displays and Disposition

Displays

General McConnell, Air Force Chief of Staff, confirmed on 21 February 1968 before the Senate Armed Services Committee that the entire B-58 fleet would be phased out by June 1970. Although there were some reprieves along the way, the final four B-58 aircraft went to the boneyard six months ahead of schedule on 16 January 1970. Most of the aircraft were eventually scrapped but seven B-58 aircraft were placed on display at various locations across the country.

55-0663. This aircraft was used by CVCFW for corporate service tests including air conditioning, environmental, sonic fatigue, stability and control, wing heaviness, and wing panel for the proposed "Dynasoar" manned orbital vehicle. The aircraft supported pod drop testing and dropped a total of 10 MB-1C pods. It was the first to drop pods from both Mach 1 and Mach 2. It made the first supersonic Mach 1 pod drop on 30

TB-58A 55-0663 was one of the 30 original YRB-58A aircraft used for service tests. It tested pod drops and conducted the first supersonic pod drop. (Grissom Air Museum)

September 1957 and the first Mach 2 pod drop and first flight above 60,000 feet on 20 December 1957.

This aircraft was delivered in place under bailment to CVCFW on 9 August 1958. It was formally delivered to the Air Force (AMC) on 29 April 1959 but retained at CVCFW. It was assigned to the 6592nd TS at Carswell AFB, TX on 1 December 1959 for R&D testing and returned to CVCFW on 12 July 1960. It made the first supersonic upper TC pod drop on 11 December 1960 flying Mach 1.6 at 45,000 feet over White Sands Missile Range. The aircraft continued supporting pod testing and dropped TC pod upper components and seven lower components (several at low altitudes). It performed low-level TC pod drops over Tonopah Test Range, NV in July 1961 to test the pod retardation system. It performed a low-level drop at Eglin AFB, FL in August 1961 to test the retardation system at sea-level (also, high-level to test water impact). The aircraft was also used by NASA for Super Sonic Transport (SST) sonic boom tests in 1962.

The aircraft was modified to serve as an operational aircraft for SAC and redesignated TB-58A on 14 March 1963 after completion of the test program. It was then assigned to the 305th BMW at Grissom AFB, IN. It was subsequently damaged by fire in the cockpit caused by an oxygen leak ignited by an electrical spark in 1969. It is currently displayed at the Grissom Air Museum in Peru, IN. It was displayed outside for 18 years along with other aircraft at the museum and subject to deterioration from the harsh Indiana weather. However, it was moved into a newly constructed dedicated indoor facility in February 2026. The building was being constructed in three phases as of 2026. The first phase was getting the aircraft housed and out of the weather, which was completed in February 2026. The next phase was to add climate controls to prevent further deterioration and allow year-round viewing of the aircraft. Finally, the building would be outfitted with educational and immersive displays. The aircraft was expected to be available for public viewing again in 2027.

55-0666. This aircraft was delivered in place under bailment to CVCFW on 20 March 1958 and used for corporate service tests. It flew demonstration flights under Operation SEVEN-UP and was modified for service as the YJ79-GE-5 engine test bed. It flew 32 minutes at Mach 2 sustained design speed with YJ79-GE-5 engines installed on 8 November 1958.

The aircraft was formally delivered to the Air Force (AMC) on 29 April 1959 but was retained at CVCFW. It was used for follow-on SEVEN-UP testing at Holloman AFB, NM during 1959. The aircraft was assigned to the 6592nd TS at Carswell AFB, TX on 30 June 1960 for R&D testing and then returned to GDFW on 31 July 1962.

It made the longest early flight test program flight of 11 hours 15 minutes covering 6,600 miles and varying altitudes from 2,000 to 30,000 feet on 16 August 1962. It was also used for a series of subsonic low-level flights to measure B-58 response to atmospheric

YRB-58A (55-0666) was used for XB-70 pilot training and committed to several test programs including the J79-GE-5 engine integration program. (Castle Air Museum)

turbulence. It was assigned to the AFFTC on 17 November 1962 for XB-70 pilot training. It was transferred to Chanute AFB, IL on 22 May 1964 for ground training. The aircraft was possibly redesignated as GRB-58A in May 1965 although there is no entry in the aircraft historical records to substantiate.

It was subsequently put on display at Chanute AFB, IL in 1967 and painted as the world record holder B-58A (61-2059) “Greased Lightning.” It was later transferred to the Octave Chanute Aerospace Museum when Chanute closed in 1993. The nose gear was damaged as the aircraft was moved due to a locked main gear. It was transferred to the Castle Air Museum in Atwater, CA after the Octave Chanute Aerospace Museum closed in 2015. It arrived on 17 August 2017 for display and restoration. It remained behind chain link safety fencing until restoration was completed in early 2025.

55-0668. This aircraft was delivered in place under bailment to CVCFW on 13 May 1958 and used for corporate service tests including BNS. It was redesignated as NB-58A on 13 January 1960 and used for testing the Hughes AN/APQ-69 SLAR, Goodyear AN/APS-73 SAR, and other special projects. It was also fitted with the J79-GE-9 engines intended for the B-58B. It was modified to serve as an operational aircraft for SAC and redesignated as TB-58A on 13 April 1964.

The aircraft was retired to the Military Aircraft Storage and Disposition Center (MASDC) on 16 January 1970. It was subsequently removed and sent to the Southwest Aerospace Museum in Fort Worth, TX just outside the Air Force Plant 4 main gate where the aircraft was produced. Originally named "Wild Child II" and "Peeping Tom" it was renamed "City of Fort Worth" during its stay at the museum. The aircraft was again moved to Meacham Airport in Fort Worth, just a short distance away, for a planned new museum facility. However, the museum plans were eventually dropped, and the aircraft was moved to Air Force Plant 4 for the 50th anniversary. It was again moved to the Lonestar Flight Museum in Galveston, TX in 1990 and then to Little Rock AFB, AR Heritage Park in 2012 where it is currently on display.

The 19th Airlift Wing vice commander speaks during the dedication ceremony for the TB-58A (55-0668) on 3 May 2013 at Little Rock AFB, AR. (U.S. Air Force)

59-2437. This aircraft was a production B-58A formally accepted by the Air Force on 26 October 1960. It was formally delivered to the Air Force on 31 October 1960 and assigned to the 43rd BMW at Carswell AFB, TX as a SAC operational aircraft. It carried the names "Firefly II" and "Rigley's Baby" at different points in its service life. The aircraft was damaged beyond repair during landing at Little Rock AFB, AR on 16 July 1968. It was permanently grounded after a main landing gear strut failure and used as a training aid. It was put on display at the Kelly AFB, TX main gate (now called Port San Antonio).

59-2458. This was a production B-58A formally accepted by the Air Force on 29 April 1961. It was formally delivered to the Air Force on 3 May 1961 and assigned to 43rd BMW as a SAC operational aircraft. Named "Cowtown Hustler", and flying under the call sign "Tall Man Five-Five", it set three world speed records on 5 March 1962 during

B-58A (59-2437) was a production B-58A assigned to the 43rd BMW as an operational aircraft. (Port San Antonio)

B-58A (59-2458) "Tall Man Five-Five" earned both the Mackay and Bendix trophies when it set three world speed records on 5 March 1962. (National Museum of the U.S. Air Force)

Operation HEAT RISE for a round-trip flight between Los Angeles and New York City and received the Mackay Trophy. It also received the Bendix Trophy for the Los Angeles to New York City outbound flight that same day. It was later reassigned to the 305th BMW. It is currently on display at the National Museum of the U.S. Air Force.

61-2059. This aircraft was a production B-58A aircraft. It was accepted by the Air Force on 16 February 1962. It was formally delivered to the Air Force on 19 February and assigned to the 305th BMW as a SAC operational aircraft on 25 March 1962. It set a world speed record on 16 October 1963 for a flight from Tokyo to London and was dubbed "Greased Lightning." It was previously named "Can Do." It is currently on display at the Strategic Air Command and Aerospace Museum in Ashland, NE.

B-58A (61-2059) "Greased Lightning" set a Tokyo to London speed record in October 1963. (Strategic Air Command and Aerospace Museum)

61-2080. This aircraft was the last production B-58A aircraft. It was accepted and delivered to the Air Force on 26 October 1962 and assigned to the 305th BMW as a SAC operational aircraft. It was retired and sent to MASDC on 6 January 1970. It was later sent to the Pima Air & Space Museum in Tucson, AZ where it is currently displayed.

Disposition

Phase-out of the B-58 fleet was ordered by SECDEF Robert McNamara on 8 December 1965. During the latter part of WWII, McNamara worked for General LeMay during LeMay's massive bombing campaign against cities in Japan. Now that he was

B-58A (61-2080) was the last production B-58A aircraft ever built. (Pima Air & Space Museum)

SECDEF, he and LeMay (now Air Force Chief of Staff) disagreed on a number of issues. He was not at all supportive of LeMay's quest for the Mach 3, XB-70, heavy bomber—judging it to be too costly. McNamara decided to let LeMay's miracle bomber stay in its bottle, asking Congress only for enough money, $220 million, to continue testing. He was also willing to take the first steps toward phasing out the B-52 and B-58 bombers, allocating less money for those planes than the Air Force had requested.[1]

He directed the phase out of 345 older B-52C through F models and all B-58 bombers by June 1970. Between 8-10 December 1965 many prominent individuals expressed serious concerns over this dramatic and unexplained cut in SAC aircraft capability. SAC objected and stated that the B-58A was now becoming a sound and effective weapon system. It was also feared by the Soviets because of its ability to evade enemy radar. SAC wanted to keep the aircraft until at least 1974 due to the declining manned bomber force. On 10 December, at a somewhat delayed press conference, McNamara announced the decision to procure a new FB-111A aircraft to replace the retiring B-52 and B-58 aircraft.[2] He anticipated procuring 210 FB-111A aircraft, which along with 255 retained B-52G and H models gave SAC a total bomber force of 465 aircraft in 1971 (reduced from 680 in 1966).

McNamara considered the FB-111A to be a more cost-effective aircraft than both the older B-52 models and the B-58, albeit not superior to those aircraft. The aircraft would be a slight modification to the F-111 to include extended wing tips, larger external fuel tanks, strengthened landing gear, improved avionics, and Short Range Attack Missile (SRAM) capability. It provided supersonic capability, low-level capability including TFR (connected to the autopilot), high speed penetration, modern avionics, and low radar cross section. Retaining 255 B-52G and H aircraft would be sufficient unless the Soviets significantly improved their defenses. In this case, the additional 210 FB-111A advanced bombers were needed.

By this point, McNamara had also concluded that the missile force was the primary weapon for the Mutually Assured Destruction (MAD) mission and that manned bombers were now merely a supplemental force. He believed that both the B-52G and H force

McNamara considered the FB-111A to be a more cost-effective aircraft than both the older B-52 models and the B-58, albeit not superior to those aircraft. (U.S. Air Force)

B-52G (57-6518) loaded with AGM-69A SRAMs in the early 1970s. The aircraft could carry a total of 20 missiles with eight in the bomb bay and six on each wing pylon. (U.S. Air Force)

and the FB-111A carrying SRAMs would be a highly effective supplemental force in case the missile force was rendered less effective than expected. He supported maintaining a manned bomber force to compel the Soviets to maintain both their present anti-bomber and missile defenses. This forced them to allocate forces in ignorance of which forces would attack where and complicated their defense strategy. He concluded this could be performed with a smaller bomber force.

On 8 March 1966, a House of Representatives subcommittee began lengthy and comprehensive hearings on U.S. Military Posture. The hearings were chaired by Congressman Mendel Rivers of South Carolina and continued through early May 1966. McNamara stated before the committee that the FB-111A range on a typical nuclear strike mission exceeded both the B-52C to F models and the B-58A. However, his analysis was based on significantly different nuclear bomb loadings which had the net effect of significantly misrepresenting the comparative capability of these aircraft. A comparative operational analysis indicated that the FB-111A had the following drawbacks:

- The FB-111A and the B-52F had relatively slight performance and range differences although the FB-111A could reach the target approximately one hour sooner.

- Although the B-58A had a slightly shorter low altitude radius it reached the target faster than the FB-111A and delivered nuclear ordnance weighing 50 percent more than the FB-111A with a higher megaton yield.
- The FB-111A could reach the target at the same time as the B-52D, even though it would fly much faster in the vicinity of the target.
- A B-58A with five nuclear weapons loaded provided a megatonnage greater than the FB-111A and could reach the target 1 hour and 23 minutes sooner.
- The FB-111A did have conventional weapons capability while the B-58A did not, but that capability was marginal at best.
- The B-52D could deliver 3.5 times the conventional weapons payload on target as the FB-111A, requiring more than three FB-111A aircraft to equal one B-52D.
- The FB-111A would require frequent refueling.

After reviewing this new analysis, the subcommittee concluded that the FB-111A was no panacea. If the old B-52 models could not be retained, though, they concluded it would be better to have something, even if less than desired, to replace them. They believed the FB-111A would be an interim aircraft until the AMSA could be made available, which was contrary to McNamara's belief that it could be a permanent B-52/B-58 replacement. Although the FB-111A could be used effectively, its range limitations and configuration prevented it from carrying the wide range of nuclear weapons required in the SAC inventory. Thus, its nuclear ordnance was severely limited in both size and megaton yield. This further meant that the FB-111A could reach only 70 percent of the targets assigned to SAC. Additionally, the committee judged that the FB-111A could not efficiently support the airborne alert role due to frequent aerial refueling requirements and human fatigue from the two-man crew. The FB-111A alone could not fulfill SAC requirements. (This may be why McNamara assumed the B-52G and H models would also be retained).

The subcommittee was also shocked to discover that the proposal to phase out the B-58 was solely an action recommended and supported by the SECDEF and not recommended or truly supported by the Air Force or the Joint Chiefs of Staff (JCS). There were about 80 B-58 aircraft still in service, and it was the only bomber aircraft capable of supersonic speeds. They were neither worn out nor obsolete and were fully capable of performing mission requirements. The committee believed it was inconceivable that a relatively new and proven weapon system developed after $3 billion of investment would be announced for elimination even before its replacement was proven and operationally available.

Every military witness strongly opposed eliminating the B-58 or at least deferring the action until the FB-111A became operational. Both former CINSACs, General Power and surprisingly LeMay, strongly recommended retaining the B-58 after 1970. During

Every military witness strongly opposed eliminating the B-58 or at least deferring the action until the FB-111A became operational. (American Aviation Historical Society)

the hearing, based on the testimony from the military advocates, McNamara agreed to delay the B-58 phase out from 1970 to 1971 so the JCS could revisit it in their next Five-Year Joint Strategic Operations Plan. Representative Lennon of South Carolina then asked McNamara if the B-52G and H models, along with the 210 FB-111A aircraft, would be an adequate bomber force. McNamara easily replied, "Yes, without question", to which Lennon said, "Mr. Secretary, I inquired of Secretary Brown of the Air Force, the Chairman of the Joint Chiefs of Staff, the individual members of the Chiefs of Staff of the several services, and I cannot find any one of them who says that they ever recommended, or that they were ever consulted, with respect to the phasing out of the B-58 by 1971. On the contrary, all of them say it is an excellent plane. It has supersonic dash that you do not find in any categories of the B-52s. They think it is a critical mistake to project the phasing out of the B-58 by 1971."

Clearly, there was no agreement from the Air Force or the JCS that the B-58 should be phased out. While they agreed to phasing out the B-52C through F models and replacing them with the FB-111A (still to be developed), they understood the B-58 phase out

SAC maintained 50 percent of the B-58 force on ground alert at both Little Rock and Bunker Hill supporting the SIOP. (National Museum of the U.S. Air Force)

would now be in 1971. This gave some time to reconsider and reverse the decision if factors changed. In 1966, the B-58 was considered an excellent aircraft with structural integrity validated through 1972. Its operational effectiveness was fully demonstrated through bombing competition performance and ORI tests. The effectiveness and reliability could be maintained through 1968 without significant modifications. Proposed ECM and BNS modifications after 1968 would extend operational effectiveness to 1972 and beyond. SAC maintained 50 percent of the B-58 force on ground alert at both Little Rock and Bunker Hill supporting the SIOP. The B-58 could also support strategic reconnaissance missions using a special LA-331 pod with cameras installed and mounted to the aircraft. It could carry five nuclear weapons to the target resulting in a larger yield than the FB-111A.

McNamara resigned on 29 February 1968 and the JCS submitted their revised force planning document to the new SECDEF, Clark Clifford, in April 1968. The document showed the B-58 still programmed through 1974. In response, on 9 January 1969, Deputy SECDEF Paul H. Nitze made minor program changes modifying all B–52 and FB–111A aircraft to carry SRAMs and the Subsonic Cruise Aircraft Decoy (SCAD).[3] All B-58 aircraft would be retained in service until SCADs became available. This substantiated keeping the B-58 in the force until 1974.

B-58A (59-2441) record holder at the boneyard. (U.S. Air Force)

However, in 1969 Melvin Laird became the new SECDEF under the Nixon administration. Laird announced a round of base closings on 27 October

1969. Both Little Rock AFB, AR and Bunker Hill AFB, IN were identified to lose their B-58A aircraft although they would remain as active bases (with new missions). Suddenly, the B-58 was back on the chopping block and again scheduled for retirement by June 1970. SAC leaders agreed with the B-58 retirements with the understanding that the B-52C through F models would be spared. But much to their surprise, Laird also subsequently ordered the retirement of all remaining B-52C through F aircraft.

SAC ultimately completed the retirement of all B-58 aircraft six months ahead of schedule. Final retirements started on 3 November 1969 when the first B-58A (59-2446) from the 43rd BMW arrived at the MASDC boneyard at Davis-Monthan AFB, AZ. They were completed on 16 January 1970 when the last two aircraft (55-0668 and 58-1007) from the 43rd BMW and the last two (55-0662 and 61-0278) from the 305th BMW arrived at MASDC. A total of 74 B-58A and eight TB-58A were retired to MASDC and seven were placed on display (two were sent to MASDC before later being put on display).

B-58A aircraft in storage awaiting disposition at Davis-Monthan. (U.S. Air Force)

In his book, *The B-58 Blunder*, Colonel George Holt Jr. (retired), a former B-58 navigator/bombardier and expert speculated on why SAC leadership eventually surrendered the B-58 fleet. According to Holt, B-58 crews gained a reputation as being a cut above the average aircrew, and this did not sit well with many in SAC. Because the Hustler was the most sophisticated and technologically advanced aircraft of its day, it required the very best crew members to fly it. We could not just volunteer for B-58s. If we were coming from B-47s or B-52s, we had to have accumulated a large number of flying hours. Additionally, we had to be selected and recommended by our wing commander to enter the B-58 program. However, that did not guarantee we would be on a B-58 crew. Pilots, especially, had to acquire a new skill set to fly the Hustler and many 'washed out' during the TF-102A (delta wing) phase of their training. I heard a rumor that some of those "washed out" pilots from the B-58 program had found their way into the Operations directorate of the Pentagon and had a role in getting rid of the B-58 and its 'Prima Donna' aircrews. Probably not true, but it highlights the fact that there were some 'sour grapes' that were not overly supportive of the B-58 program.

Holt continued, in the fall of 1969, while assigned to the Future Strategic Forces, Plans Division, at the Pentagon, I received a phone call from Lt. Colonel Al Dugard, who was working at SAC, in Omaha, Nebraska. "George, I need to give you a heads up. This came as a complete surprise to me. Our Director of Plans, Major General Sherman Martin is preparing to head for the Pentagon to convince General Ryan (Air Force, Chief of Staff) that the Air Force should trade off both B-58 Wings in exchange for keeping four older B-52 Wings that the Secretary of Defense wants to get rid of." "Al, you've got to be kidding. I thought the Air Force had won the B-58 battle." "You're right, but SAC changed its mind, and they are now willing to give up the 58s. I and other B-58 drivers here at SAC, were never even consulted. We had no input to the decision." "What was SAC's reasoning?" "George, just between you and me, I'm convinced it was a political move, within SAC. It will preserve four Wing Commander positions for four Colonels; and of course, the B-52 faction is in the majority here at SAC headquarters. Even our Deputy Chief of Staff for Operations, Major General Selmon Wells, always a supporter of the B-58s had no influence on the decision."

Well, did I get a surprise. General Martin decided to plead his case through the Operations side of the house. He came in through the back door, or I should say, the 'basement door' to the Pentagon. That is where the Operations guys hung out. Anyhow, he got the Ops people to agree with him and by the time I found out he had already briefed General Ryan and got Ryan's agreement to get rid of the B-58s. Unbelievable. Therefore, I decided to fight that decision. The General had made a number of assertions about the B-58s and the B-52s that simply were not true. For the next couple of days, I worked 24/7 to develop a position paper and coordinate it through a number of Air Staff directorates. It contained the following major points:

Warfighting capability: For its role in the SIOP the B-58, when compared to the B-52, was shown to be more survivable against Soviet air defenses, and more capable of reaching and destroying its assigned targets.

Survivability Assessment: 1. The B-58 flew much faster than the B-52 at low-level. 2. The B-58 had a much smaller radar cross section, whereas the B-52's much larger radar signature made the B-52 more likely to be detected by enemy radars. 3. I had the Studies and Analysis Directorate at the Pentagon run computer simulations comparing the B-58 with the B-52. Their conclusion; the B-58's speed advantage and its very low radar signature gave the Hustler a higher probability of evading detection by enemy radars. This held true even when the B-58 was programmed to fly at a higher altitude than the B-52 during low-level penetration to the target. In comparing radar signature differences, I remember one of the evaluators saying, "The difference between the B-58 and the B-52 was like comparing a postage stamp to a barn door." 4. My own operational experience in flying low-level missions in the B-58 confirmed this finding. Quite often, the radar bomb scoring sites in the U.S. had difficulty picking us up on radar as we flew inbound to the target. I would give them our distance to the target so they could confirm that they were tracking us on radar. If we were skimming the treetops, I would ask Al Dugard, our pilot, to pop up a little in altitude so they could find us. I wanted to make sure we got credit for the bomb run.

Cost Comparison: One of the most egregious claims often repeated by many in SAC and others was, "For the cost of operating two B-58 wings, we can operate six B-52

The B-58's speed advantage and its very low radar signature gave the Hustler a higher probability of evading detection by enemy radars at both high and low levels. (American Aviation Historical Society)

wings." This sounds like the B-58 was a very costly aircraft to operate, but when a cost comparison is made aircraft to aircraft, we find just the opposite to be the case. Each B-58 actually costs less to operate than a B-52. This held true even when the higher B-58 air refueling cost was thrown into the mix. How was that possible? First, the two B-58 wings each had 39 aircraft assigned, whereas the six B-52 wings had only 15 aircraft assigned. Therefore, the cost comparison, to begin with, should have been 78 B-58s compared to 90 B-52s.

Secondly, all the B-58s were located on only two bases while the six B-52 wings were spread out on six separate bases. The reason SAC spread out the B-52s to so many bases, with a normal complement of 15 aircraft at each base, was to ensure survivability in case of an enemy attack. It took twice as long for a B-52 to get off the ground compared to the B-58. Therefore, dispersal to many bases for the B-52s was more critical to their survival. But that contributed to driving up the cost for operating the B-52s. This meant the B-58s actually incurred less base support and other associated operational, maintenance, and logistical support costs, due to economy of scale.

Thirdly, an Air Force report estimated the annual operating costs of strategic bombers, including the associated cost for KC-135 tanker refueling support for both the B-52 and the B-58. It showed that the B-58, on an airplane-to-airplane basis, was less costly to operate. The estimated average annual operating cost for a single B-52 was $1.42 million, while the comparable cost for a B-58 was $1.21 million.

According to Holt, although maintenance costs for the Hustler may have been higher, the B-58 had a definite advantage in lower personnel costs, requiring 50 percent fewer crew members per aircraft than the B-52. Now if we do the cost comparison, on an airplane-by-airplane basis we find that 78 B-58s at a cost of $1.21 million each, equals $94.4 million whereas 90 B-52s at a cost of $1.42 million each, equals $127.8 million. Following the above math, it shows that for $94.4 million (the cost of operating the B-58s) we could have operated only 66 B-52s. The above claim could more accurately be stated as; For the cost of operating 66 B-52s, we can operate 78 B-58s.

It was also commonly believed that the B-58 was an expensive aircraft to buy. While it is true that the aircraft that were delivered cost more than the B-52, for example, the higher price was due largely to the small number of aircraft bought. B-58 production stopped at slightly more than 100 aircraft. Contractor estimates show that the B-58 and B-52 cost per aircraft evened out at about 150 aircraft. At aircraft number 200 the contractors believed the flyaway cost of the B-58 would have been $1.1 million (1959) less than the B-52.[4]

Other B-58 Advantages: Along with its ability to rapidly start engines, taxi, and take-off, there were other B-58 advantages like a much better synchronous bombing capability,

Although maintenance costs for the Hustler may have been higher, the B-58 had a definite advantage in lower personnel costs, requiring 50 percent fewer crew members per aircraft than the B-52. (American Aviation Historical Society)

a better capability to withstand low-level turbulence, and its ability to fly silent. Holt said, I had flown numerous low-level navigation missions in the B-58. On more than one occasion when we would report light turbulence on a low-level mission, B-52s in front, or behind us, would report moderate to heavy turbulence and decide to abort the low-level mission and climb back to high altitude. Unlike the B-52, the B-58 was quite adept at flying through low-level turbulence. It had a very strong body, and with its short non-bendable wing, there was no problem with wing oscillations like the B-52 experienced.

Then came the bad news, Holt wrote, "Well done Major Holt, and I agree with everything in your position paper, but unfortunately, it's too late." "What do you mean, sir?" "General Ryan met with Secretary Laird just this morning and gave him his decision to trade off the two B-58 wings to save the four B-52 wings. And of course, Secretary Laird approved that decision." "Is there no way of turning this around?" "I'm afraid not. You have to realize that this decision did not originate with General Ryan. It was a decision made by General Bruce Holloway, Commander in Chief, of Strategic Air Command. General Ryan could not go against the desires of one of his most important operational commanders." "But it was a wrong decision." "I hear you, George, ... I'm afraid it's a done deal and we'll have to live with it." I was speechless. I picked up my paperwork, saluted smartly, and left his office. I lost the battle.

In his book, Holt also countered the following common myths with facts about the B-58 Hustler.

Myth 1: "In late 1965, Secretary McNamara ordered the B-58's retirement by 1970. Despite efforts of the Air Force to earn a reprieve, the phase-out proceeded on schedule." Fact: Not true. The phase-out did not proceed on schedule. It was interrupted twice. Once by McNamara in 1966 when he extended the phase-out to 1971. He also gave the Air Force, through the JCS, the opportunity to provide recommendations to retain the Hustler. In April 1968, the Joint Chiefs of Staff submitted their Force Planning document, Part VI of JSOP-70-77. It showed the B-58 still programmed to be in the bomber force beyond 1974. In addition, on 9 January 1969, Deputy SECDEF Paul Nitze made a program change to retain the B-58s in service until SCADs became available for B-52s. This effectively pushed the B-58 phase-out date to at least 1974. The only thing that cancelled that 1974 date was the blunder by SAC when, in the latter part of 1969, they offered to trade off 78 B-58s for 90 older model B-52s.

Myth 2: "Originally intended to fly at high altitudes and high speeds to avoid Soviet fighters, the introduction of highly accurate Soviet SAMs forced the B-58 into a low-level penetration role that severely limited its range and strategic value." Fact: Soviet SAMs forced both the B-52 and the B-58 into a low-level penetration role. For that matter, even the B-47 was forced into that role before it was retired. However, the B-58 adapted to the low-level mission quite well. In September 1959, even before the B-58 entered the operational force, it flew three low altitude penetration flights. One four-hour mission on 18 September included a two-hour flight at 0.92 Mach and 500 feet or less AGL. One portion was as low as [and below] 200 feet AGL. Its range was more than adequate to reach its assigned targets in the Soviet Union. In fact, its strategic value actually increased when forced into the low-level penetration mode, because compared to the B-47 and the B-52, it was much faster, less susceptible to low-level turbulence, and had the best BNS. Its ability to fly silent made it virtually undetectable by enemy radars.

Myth 3: "The B-58 cost three times as much to operate as the B-52." Fact: As previously shown, when a cost comparison is made aircraft by aircraft, we find just the opposite to be true. Each B-58 actually costs less to operate than a B-52.

Myth 4: "Although the aircrews swore by the B-58, money and mission limitations led to it being phased out of the inventory after only 10 years of service." Fact: It is true that aircrews swore by the B-58. They loved the airplane and its capabilities. Money issues were not the problem. The B-58s were bought and paid for and the B-58s operating costs turned out to be less than the B-52s. Moreover, it had no mission limitations. All B-58s were well capable of reaching their assigned SIOP targets. For its SIOP role the average nuclear bomb carrying capability was actually greater than the B-52.

Most aircrews swore by the B-58 and the loved the airplane and its capabilities. (American Aviation Historical Society)

Myth 5: "Since the B-58 structure incurred significant fatigue damage when flying at low-level, and since the new bomber had no TFR, extensive modifications would be needed to permit effective low-level penetration." Fact: Like the B-47 and the B-52, the B-58 started showing signs of structural fatigue after numerous hours of low-level flying. However, in 1967 all B-58 aircraft underwent extensive inspection and repair of all primary and secondary structures.[5] The B-58 was much less susceptible to low-level turbulence than the B-52. The strong delta wing and short profile of the Hustler made it relatively free from the oscillation effects of low-level turbulence experienced by the B-52. It is true that the B-58 had no TFR, but because of its high speed, low radar signature, and ability to withstand heavy turbulence at low altitude, it was very effective at low-level penetration to its targets. Most historians are probably unaware that by 1969 a highly accurate Radar Altimeter was installed allowing the B-58 to fly even lower on its penetration to enemy targets.

Myth 6: "Some SAC brass disliked the B-58 because of its high accident rate." Fact: Compared to other aircraft, the 'operational' accident rate was not that bad and quite different from the oft-quoted phrase; "Of the 116 built, 26 were destroyed for an accident rate of 22%." Actually, of the 116 built, 24 (not 26) were destroyed before the Hustler was finally phased out. However, seven of those accidents occurred before the Hustler ever entered the Air Force inventory to become operational. A more meaningful accident rate would read as follows; "Of the 116 built, 17 were destroyed after the B-58 became operational for an accident rate of 15%." Of the 17 aircraft that were lost after the B-58 became operational, two were lost at Paris Air Shows (one was performing an aerobatic maneuver and the other impacted the ground by flying too low and hitting runway approach lights.) Another was lost in the taxiing incident at Bunker Hill AFB. Therefore, if we choose to discount those three aircraft, and concentrate only on aircraft lost while flying operational missions we have a total of 14 B-58s lost over a span of 10 years. An average loss of 1.4 aircraft per year is not a bad accident rate, at all, when compared to other combat aircraft. The Mach 3, SR-71 (and variants) had an accident rate twice that of the Hustler, and the B-47 experienced 28 fatal accidents and 63 deaths in just one year (1957).

Myth 7: "The B-58 had several design flaws that shortened its life as a SAC bomber." Fact: By the time the B-58 completed RDT&E (Research, Development, Test and Evaluation) almost all "design flaws" had been worked out. Finding 'design flaws' led to 'design innovation.' There were many design improvements due to innovative engineering breakthroughs.

Myth 8: "The B-58's planned production run was reduced because of the high cost per unit, a small payload, a mission profile different from its design concept, and, in the 1960's, a Secretary of Defense that downplayed the role of the bomber." Fact: R&D costs are sunk costs in the purchase of any weapon system. Therefore, if the planned production run is cut in half, then the total system unit cost has to increase. It had a tremendous payload for its wartime SIOP mission—better than the average nuclear payload of the B-52. Although its low-level mission profile was different from its initial design concept, it was amazingly well adapted to this new profile and there was no other bomber as capable as the Hustler in this role. The only factual part of this myth is that McNamara, as SECDEF, consistently downplayed the role of the bomber, especially the B-58.

Myth 9: Designed for supersonic, high-altitude penetration, the B-58 was limited in range, payload, and growth potential for the addition of advanced radar and other electronic equipment. Fact: Both the B-58 and the B-52 required air refueling for their wartime SIOP missions, but this did not limit the range for the B-58. Its range, with a single refueling, was adequate to attack all of its assigned SIOP targets and its low altitude speed was much faster than the B-52. It is true that the B-58 had no TFR, but because of its speed, low radar signature, and ability to withstand heavy turbulence at low altitude, it was actually more effective at low-level penetrations to its targets. In the 1968-69 period, a number of advanced improvements were made to the Hustler, including an improved version of the AN/ASQ-42 BNS and the installation of a highly accurate Radar Altimeter, which allowed the B-58 to fly even lower on its penetration to enemy targets.

B-58A range, with a single refueling, was adequate to attack all of its assigned SIOP targets. (National Museum of the U.S. Air Force)

Myth 10: The B-58 was a nightmare for maintenance. Fact: There is some truth to this, but it must be placed in perspective. First, the B-58 was a very tight airplane, meaning there was no wasted space. There was lots of specialized test equipment and access to anything needing repair required removal of panels. The B-58 experienced excessive down time after a mission, as discrepancies had to be cleared before the next flight. Before any maintenance could begin on the aircraft, a ground air conditioning unit had to be hooked up. Then there were the 'Hangar Queens' that had numerous repeatable maintenance problems that no one could figure out. Quite often, those problems were associated with the BNS. However, starting in May 1968 the entire fleet received an improved version of the AN/ASQ-42 BNS, replacing old vacuum-tube technology with [more reliable] transistorized components. Some maintenance personnel said they "hated working on this airplane" but in almost the same breath, they would say, "they wouldn't trade it for the world."

Myth 11: The B-58 was never easy to fly. Fact: I've found just the opposite to be true. Among all the pilots I've known in my three years of flying in the B-58, none thought the B-58 was hard to fly. In fact, they thought it was the smoothest airplane they ever flew.

Especially those who have flown fixed wing (B-47s and B-52s) as well as other Delta wing (F-102/F-106s) aircraft. This includes Colonel Al Dugard who was a B-58 Combat Crew Training School instructor pilot. He said, "All deltas give a smoother ride than other aircraft. Responsiveness to controls is instantaneous as you don't wait for a wing to respond to control movement. Stability of flight is because of the delta wing. Formation flight i.e., air refueling was much easier due to the stable platform."

Landings were made at high speed because the B-58 did not have flaps or boards to decelerate. It used a drag chute to slow it down like both the B-47 and B-52. (American Aviation Historical Society)

Myth 12: It required a much higher AoA than a conventional aircraft, up to 9.4 degrees at Mach 0.5 and low altitudes. Fact: The B-58 flew at much higher speeds than Mach 0.5. At cruise altitude the B-58 would fly at Mach 0.91. Even low-level training missions were never flown below Mach 0.70. In both cases, the AoA was much lower. Cruising at optimum altitude the AoA, regardless of gross weight, would be about 6 degrees and at low-level it would run about 5 degrees.

Myth 13: The plane had very unusual takeoff requirements, with a 14-degree AoA needed for the rotation at about 203.5 knots. This poor takeoff performance was also evident with the high landing speed that necessitated a drogue parachute for braking. Fact: The inference is that the B-58 had poor takeoff and landing performance when just the opposite was true. A 14-degree AoA is normal for a delta wing aircraft as it generated immediate lift, and with the power of four engines in full afterburner, the airspeed quickly accelerated to 425 knots climb speed. Landings were made at high speed because the B-58 did not have flaps or boards to decelerate. Yes, it used a drag chute upon touch down, but that was not peculiar to the B-58.

Myth 14: Another problem pilots faced was called "fuel stacking" and took place when the B-58 accelerated or decelerated. It was due to fuel moving in the tanks and causing sudden changes in the CG. This could cause the aircraft to pitch or bank and subsequently lose control. Fact: This may have occurred in one of the early flight tests but "fuel stacking" was never a problem with the operational fleet of B-58s. Fuel was shifted effortlessly [and automatically] when going into or coming out of supersonic flight.

Myth 15: The B-58 was very difficult to safely recover from the loss of an engine at supersonic cruise due to differential thrust. Fact: The installation of a "Triple Redundant Yaw Damper" system allowed safe recovery when an outboard engine was lost, even while flying at Mach 2.

Appendix A: Specifications

YRB-58A Hustler

The YRB-58A performed R&D testing and training duties and some were later converted to SAC operational aircraft. (U.S. Air Force)

Aircraft Specifications					
Type:	B-58	Series:	YRB-58A	Name:	Hustler
Wingspan:	56.8 ft	Length:	96.8 ft	Height:	31.4 ft
Empty Weight:	51,061 pounds	Combat Weight:	107,250 pounds	Gross Weight:	158,000 pounds
Combat Radius:	4,500 miles	Combat Ceiling:	59,000 ft	Service Ceiling:	59,400 ft
Cruise Speed:	626 mph	Max Speed:	1,320 mph	Max Payload:	19,450 pounds
OEM:	Convair	Produced:	30	SAC Inventory:	8
First Flight:	11-Nov-56	First Delivery:	14-Feb-58	Phase Out:	22-May-64
Missions:	Strategic Medium Bomber R&D Test Aircraft				
Tail Number(s):	55-0660 to 55-0672; 58-1007 to 58-1023				
Propulsion:	Four General Electric J79-GE-5 (or-5A) jet engines with 9,370 (9,700) pounds normal thrust, 10,000 (10,300) pounds military thrust, and 15,600 pounds each with afterburner. One engine mounted on each of four wing pylons.				
Accommodations:	Total 3. Pilot, navigator/bombardier, and DSO. The navigator/bombardier operated the BNS. The DSO operated the ECM and the remote-controlled tail gun turret and performed duties as a performance engineer.				
Payload:	Up to 19,450 pounds of bombs including one W39Y1-1 in MB-1C or one BA53 in TC pod; plus one radar aimed 20-mm T-171E-3 Vulcan rotary cannon with 1,040 rounds of ammunition.				
Comments:	Specifications based on B/RB-58A Standard Aircraft Characteristics with J79-GE-5 engines and MB-1C pod, dated 10-Jul-59. Empty weight is 51,061 pounds without pod and 53,581 pounds with pod. OEM is Convair Division of General Dynamics Corporation. Original plans called for testing with P&W J57-P-15 engines installed on the first nine aircraft, but aircraft delays allowed the first 13 YRB-58A aircraft to be fitted with J79-GE-1 engines with 8,900 pounds normal thrust, 9,300 pounds military thrust, and 14,350 pounds thrust each with afterburner. YRB-58A (55-0666) was fitted as the testbed for -5 engines and the -5 was cut into production beginning with the 14th aircraft (58-1007). Total 8 YRB-58A transferred to SAC upon activation of 43rd BMW. Eleven YRB-58A (58-1009 to 58-1011; 58-1013 to 58-1016; 58-1018 to 58-1021) aircraft were converted to B-58A and eight (55-0661 to 55-0663; 55-0668; 55-0670 to 55-0672; 58-1007) were converted to TB-58A for SAC operational use.				

B-58A Hustler

The B-58A was the only supersonic bomber ever produced capable of flying at Mach 2+ speeds. (U.S. Air Force)

Aircraft Specifications

Type:	B-58	Series:	B-58A	Name:	Hustler
Wingspan:	56.8 ft	Length:	96.8 ft	Height:	31.4 ft
Empty Weight:	55,560 pounds	Combat Weight:	82,595 pounds	Gross Weight:	163,000 pounds
Combat Radius:	4,920 miles	Combat Ceiling:	63,080 ft	Service Ceiling:	63,500 ft
Cruise Speed:	579 mph	Max Speed:	1,320 mph	Max Payload:	19,450 pounds
OEM:	Convair	Produced:	97	SAC Inventory:	97
First Flight:	Sep-59	First Delivery:	1-Dec-59	Phase Out:	16-Jan-70
Missions:	Strategic Medium Bomber				
Tail Number(s):	58-1009 to 58-1011; 58-1013 to 58-1016; 58-1018 to 58-1021; 59-2428 to 59-2463; 60-1110 to 60-1129; 61-2051 to 61-2080				
Propulsion:	Four General Electric J79-GE-5A or -5B jet engines with 9,700 pounds normal thrust, 10,300 pounds military thrust, and 15,600 pounds thrust each with afterburner. One engine mounted on each of four wing pylons.				
Accommodations:	Total 3. Pilot, navigator/bombardier, and DSO. The navigator/bombardier operated the BNS. The DSO operated the ECM and the remote-controlled tail gun turret and performed duties as a performance engineer.				
Payload:	Up to 19,450 pounds of bombs including four pylon-mounted Mk-43 or B61 nuclear bombs and one W39Y1-1 in MB-1C or one BA53 in TC pod; plus one 20-mm T-171E-3 Vulcan rotary cannon with 1,200 rounds of ammunition.				
Comments:	Specifications based on basic, high altitude, refueled mission with MB-1C pod carrying one W39Y1-1 warhead. Empty weight is 55,560 pounds without pod and 64,115 pounds with pod. The aircraft could be refueled inflight to a gross weight of 176,890 pounds. Total of 34 B-58A aircraft modified in 1961-62 to take-off with 176,890 pounds. Remaining aircraft limited to 163,00 pounds. Combat radius 4,920 miles with refueling. Cruise speed is 579 mph for refueled mission and 610 mph for unrefueled mission. OEM is Convair Division of General Dynamics Corporation. Engines upgraded to J79-GE-5C to accommodate cartridge self-starter system installation. Eleven B-58A (58-1009 to 58-1011; 58-1013 to 58-1016; 58-1018 to 58-1021) aircraft were converted from YRB-58A. The remaining 86 B-58A aircraft were new production builds.				

TB-58A Hustler

The TB-58A was a trainer version required to orient new pilots to the challenging flight characteristics of the B-58A. (U.S. Air Force)

Aircraft Specifications					
Type:	B-58	Series:	TB-58A	Name:	Hustler
Wingspan:	56.8 ft	Length:	96.8 ft	Height:	31.4 ft
Empty Weight:	55,560 pounds	Combat Weight:	82,595 pounds	Gross Weight:	163,000 pounds
Combat Radius:	4,920 miles	Combat Ceiling:	63,080 ft	Service Ceiling:	63,500 ft
Cruise Speed:	579 mph	Max Speed:	1,319 mph	Max Payload:	N/A
OEM:	Convair	Produced:	8	SAC Inventory:	8
First Flight:	10-May-60	First Delivery:	2-Aug-60	Phase Out:	16-Jan-70
Missions:	Strategic Medium Bomber Trainer				
Tail Number(s):	55-0661 to 55-0663; 55-0668; 55-0670 to 55-0672; 58-1007				
Propulsion:	Four General Electric J79-GE-5A or -5B jet engines with 9,700 pounds normal thrust, 10,300 pounds military thrust, and 15,600 pounds each with afterburner. One engine mounted on each of four wing pylons.				
Accommodations:	Total 3. Pilot, Instructor Pilot, and DSO.				
Payload:	N/A				
Comments:	Specifications based on basic, high altitude, refueled mission with MB-1C pod. Empty weight is 55,560 pounds without pod and 64,115 pounds with pod. The aircraft could be refueled inflight to a gross weight of 176,890 pounds. Total of two TB-58A aircraft modified in 1961-62 to take-off with 176,890 pounds. Combat radius 4,920 miles with refueling. Cruise speed is 579 mph for refueled mission and 610 mph for unrefueled mission. OEM is Convair Division of General Dynamics Corporation. All eight TB-58A aircraft produced were converted from YRB-58A. Recognizable by additional cockpit windows and absence of tail gun. The aircraft featured dual pilot seating. Most navigator/bombardier's equipment removed, position closed, and a new escape hatch installed over a newly designed second station. All of the tactical equipment was removed and there was no autopilot, no primary navigation system, no bombing system, no defensive ECM system, and no active defense system. SACseat retained.				

Appendix B: Colors and Markings

The first R&D aircraft YRB-58A (55-0660) in its original paint scheme. It featured a white fuselage underside, and the top of the nose was painted black. The tips of the nacelles were red. The "Convair B-58" marking was painted in gold on both side of the fuselage. The national emblem was on both sides of the forward fuselage and the top of the left wing. The "U.S. AIR FORCE" marking was on both sides of the fuselage and the "USAF" marking was on the top of the right wing. (U.S. Air Force)

YRB-58A (55-0660) repainted with the same scheme used on follow-on R&D aircraft. The underside was now red, and this was carried to the top of the fuselage leaving a white delta shape. The top of the nose was painted black. The tail was also painted red, and a white delta was applied (like on the forward fuselage). The wingtips were painted red with white trim. The red was removed from the nacelles. All other markings remained the same. Some pods were painted in various versions of red, red/black, or red/white. (U.S. Air Force)

Production B-58A aircraft had a relatively bland paint scheme. The aircraft was completely bare metal except the black nose radome and a small section of black paint in front of the pilot's windows to reduce glare. All other markings remained the same except the "Convair B-58" marking was removed. (U.S. Air Force)

Units eventually began adding unit patches as shown on the forward fuselage. Initially, these were without the banner shown. Different tail markings were also added. (U.S. Air Force)

The SAC Shield and blue "Milky Way" banner became mandatory for all SAC aircraft including the B-58. The banner was also applied on the right side of the fuselage under the unit patch. Notice the tail number on the nose gear door is now painted red instead of the previous black. (American Aviation Historical Society)

B-58A (59-2437) displays its 305th BMW unit patch and unique tail marking. (Port San Antonio)

YRB-58A (55-0662) also sported a unique tail marking with a black delta outlined in white on a red tail. The delta also contained the AFFTC unit patch. Also, note the unit patch on the forward fuselage with no banner. (American Aviation Historical Society)

The first TB-58A (55-0670) carried a unique paint scheme similar to the initial R&D aircraft. The underside of the nose was black, and this was carried to the top of the nose and under the windows leaving an exaggerated white delta shape (with a thin red border on the aft edge). The top was painted red and the area around the windows and the forward landing gear doors were painted white. The tail was also painted red, and a white delta was applied. The wingtips were painted white with red trim. A red and white delta was painted on each nacelle. All other markings remained the same. This particular pod was also painted red with a white delta.

Appendix C: Tail Numbers

Tail #	CN	Lineage	Status	Last Unit	Date	Comments
55-0660	1	YRB-58A NB-58A YRB-58A GRB-58A	Retired	CVCFW	15-Mar-60	First flight 11-Nov-56. Used by CVCFW for corporate service tests. Logged 150 total flights totaling 257 hours and 30 minutes, including 28 hours and 23 minutes at supersonic speeds. First B-58 to fly at night, first to fly supersonic on 30-Dec-56, first flight with an Air Force pilot and reached Mach 1.35 on 22-Jan-57, and first to fly at Mach 2 on 29-Jun-57. Formally accepted by the Air Force at CVCFW on 29-Nov-57. Flew 11 missions for 10 consecutive days to demonstrate durability during Operation WHIRLWIND in Feb-58. Flew at Mach 1.2 for 1 hour and 31 minutes, which was probably the longest sustained supersonic flight at that time. Conducted first missile launch from a B-58 flying out of Eglin AFB, FL on 5-Sep-58. Delivered in place under bailment to CVCFW as YRB-58A on 11-Nov-58 but incorrectly documented as B-58A. MDS corrected to YRB-58A on 22-Dec-58. Formally delivered to the Air Force (AMC) on 29-Apr-59 but retained at CVCFW. To 6592nd TS at Carswell AFB, TX on 24-Dec-59 for R&D testing and redesignated NB-58A. Fuel line surge tests in late 1959. ALBM transport in 1959/60. To CVCFW on 24-Feb-60 and redesignated YRB-58A. To Kelly AFB, TX on 15-Mar-60 for ground trainer. Possibly redesignated GRB-58A (no entry in aircraft record cards). Later scrapped in-place. Named "Old Grandpappy."
55-0661	2	YRB-58A TB-58A	Retired	305th BMW	9-Jan-70	Used by CVCFW for corporate service tests. Made first flight on 16-Feb-57 with MB-1C pod attached reaching Mach 1.15. Formally accepted by the Air Force at CVCFW on 29-Nov-57. Delivered in place under bailment to CVCFW as YRB-58A on 16-Feb-58 but incorrectly documented as B-58A. Refueled from a KC-135 tanker for the first time on 11-Jun-58. Formally delivered to the Air Force (AMC) and incorrectly documented as YB-58 and then RB-58A on 29-Aug-58. First aircraft assigned to 6592nd TS at Carswell AFB, TX on 29-Aug-58 for R&D testing. MDS corrected to YRB-58A on 22-Dec-58. To CVCFW on 20-Aug-60. First crew escape capsule static test from navigator/bombardier's station in Mar-61 at CVCFW. To the AFFTC at Edwards AFB, CA on 27-May-61 for flight test of the escape capsule in 1961-62. First human ejection from the escape capsule on 28-Feb-62. To GDFW on 30-Sep-62. Modified to operational aircraft for SAC,

Tail #	CN	Lineage	Status	Last Unit	Date	Comments
						redesignated TB-58A on 21-May-63. Assigned to 305th BMW. Named "Mach-In-Boid." Scrapped on 13-Jul-77.
55-0662	3	YRB-58A NB-58A YRB-58A TB-58A	Retired	305th BMW	16-Jan-70	First flight 6-May-57. Used by CVCFW for stability and control testing and corporate service test of BNS and autopilot. Used for Doppler Radar and Radar Altimeter tests. First MB-1C pod drop on 5-Jun-57. Formally accepted by the Air Force at CVCFW on 29-Nov-57. Delivered in place under bailment to CVCFW as YRB-58A on 1-May-58 but incorrectly documented as B-58A. First aircraft to blow all tires on landing on 30-Aug-58 and later used for frangible wheel testing. MDS corrected to YRB-58A on 22-Dec-58. First aircraft to complete Category I testing on 25-Apr-59. Formally delivered to the Air Force (AMC) on 29-Apr-59 but retained at CVCFW. To the AFFTC on 24-Sep-59 and redesignated NB-58A. Used as GE YJ93 engine test bed for XB-70 and XF-108. To GDFW on 21-Jun-62 and redesignated YRB-58A. Modified to operational aircraft for SAC, redesignated TB-58A on 14-May-63. To the AFFTC for XB-70 chase plane on 15-May-63. Later assigned to 305th BMW. Set record for 256 sorties without late or missed takeoff. Scrapped on 21-Jul-77.
55-0663	4	YRB-58A TB-58A	Display	305th BMW		First flight 12-Aug-57. Used by CVCFW for corporate service tests including air conditioning, environmental, sonic fatigue, stability and control, wing heaviness, and wing panel for the proposed "Dynasoar" manned orbital vehicle. First aircraft to drop pods from both Mach 1 and Mach 2. First supersonic Mach 1 pod drop on 30-Sep-57; first Mach 2 pod drop and first flight above 60,000 feet on 20-Dec-57. Formally accepted by the Air Force at CVCFW on 29-Nov-57. Delivered in place under bailment to CVCFW as YRB-58A on 9-Aug-58 but incorrectly documented as B-58A. MDS corrected to YRB-58A on 22-Dec-58. Formally delivered to the Air Force (AMC) on 29-Apr-59 but retained at CVCFW. To 6592nd TS at Carswell AFB, TX on 1-Dec-59 for R&D testing. To CVCFW on 12-Jul-60. First low-level TCP upper component drop over Tonopah Test Range, NV on 19-Nov-60. First supersonic TCP upper component drop over White Sands Missile Range on 11-Dec-60; first drop of both upper and lower components at Mach 2.0 in Apr-61. Performed low-level TC pod drops over Tonopah

Tail #	CN	Lineage	Status	Last Unit	Date	Comments
						Test Range, NV in Jun-61 to test pod retardation system. Performed low-level drop at Eglin AFB, FL in Aug-61 to test retardation system at sea-level (also, high-level to test water impact). First low-level upper component drop and first complete TC pod by Air Force crew at Tonopah in Aug-61. Dropped total of 10 MB pods, TC pod upper components, and seven lower components (several at low altitudes). Used by NASA for SST sonic boom and landing approach tests in 1962. Modified to operational aircraft for SAC, redesignated TB-58A on 14-Mar-63. Assigned to 305th BMW. Damaged by fire in the cockpit in 1969. Displayed at Grissom Air Museum in Peru, IN.
55-0664	5	YRB-58A	Destroyed	CVCFW	7-Nov-59	First flight 30-Nov-57. Delivered in place under bailment to CVCFW as YRB-58A on 14-Nov-58 but incorrectly documented as B-58A. MDS corrected to YRB-58A on 22-Dec-58. Formally delivered to the Air Force (AMC) on 29-Apr-59 but retained at CVCFW. Used by CVCFW for corporate service tests. Served as air loads data test aircraft. Used to establish B-58 inflight refueling envelopes from 27-Jun-58 to 17-Mar-59. Large "5" marked on empennage. Crashed over Duncan, Ok three miles south of Lawton, OK during flight test on 7-Nov-59 with Convair crew of two onboard after disintegrating at high altitude. Cause unknown. Test flight to simulate outboard engine failure at Mach 2. Total 2 fatalities/0 survivors.
55-0665	6	YRB-58A JB-58A NB-58A GRB-58A	Retired	AFFTC		First flight 28-Sep-57. Formally accepted by the Air Force at CVCFW on 29-Nov-57. Delivered in place under bailment to CVCFW as YRB-58A on 11-Feb-58 but incorrectly documented as B-58A. Used by CVCFW for corporate service tests. First aircraft formally delivered to the Air Force (AMC) on 14-Feb-58 and assigned to the AFFTC on 15-Feb for Category II testing under ARDC. MDS corrected to YRB-58A on 22-Dec-58 (incorrectly documented as YB-58A during much of its time at Edwards). Modified in 1959 to test the Hughes AN/ASG-18 Fire Control Radar system and GAR-9/AIM-47 missile planned for the F-108 and YF-12A. Redesignated as JB-58A on 12-Sep-59 and then NB-58A on 30-Nov-59. Flew a series of subsonic, low-level flights to measure response to turbulence at altitudes between 500 and 1,500 feet in 1961. Mods removed after test, except long nose, and the aircraft was towed to the photo

Tail #	CN	Lineage	Status	Last Unit	Date	Comments
						range at Edwards for ground trainer on 24-Feb-64. Possibly redesignated GRB-58A (no entry in aircraft record cards). Currently listed in desert storage area on the photo range at Air Force Flight Test Museum, Edwards AFB, CA. Named "Snoopy I."
55-0666	7	YRB-58A GRB-58A	Display	AFFTC		First flight 20-Mar-58. Delivered in place under bailment to CVCFW as YRB-58A on 20-Mar-58 but incorrectly documented as B-58A. Used by CVCFW for corporate service tests. Flew demonstration flights under Operation SEVEN-UP. Modified for YJ79-GE-5 engine test bed. First flew with YJ79-GE-5 engines in April 1958 and flew 32 minutes at sustained Mach 2 on 8-Nov-58. MDS corrected to YRB-58A on 22-Dec-58. Formally accepted by the Air Force at CVCFW on 20-Mar-59. Formally delivered to the Air Force (AMC) on 29-Apr-59 but retained at CVCFW. Used for follow-on SEVEN-UP testing at Holloman AFB, NM during 1959. To 6592nd TS at Carswell AFB, TX on 30-Jun-60 for R&D testing. To GDFW on 31-Jul-62. Made the longest early flight test of 11 hours and 15 minutes covering 6,600 miles and varying altitudes from 2,000 to 30,000 feet on 16-Aug-62. Also used for a series of subsonic low-level flights to measure B-58 response to atmospheric turbulence. To the AFFTC on 17-Nov-62 for XB-70 pilot training. To Chanute AFB, IL on 22-May-64 for ground training. Possibly redesignated GRB-58A in May-65 (no entry in aircraft record cards). To display at Chanute AFB, IL in 1967 (painted as 61-2059). To Octave Chanute Aerospace Museum when Chanute closed in 1993. Nose gear was damaged as aircraft was moved due to locked main gear. Moved to Castle Air Museum in Atwater, CA after Chanute Museum closed in 2015. Arrived 17-Aug-17 and placed on display behind chain link safety fence until restoration was completed in 2025.
55-0667	8	YRB-58A	Destroyed	CVCFW	4-Jun-60	First flight 14-Dec-57. Formally accepted by the Air Force at CVCFW on 14-Dec-57. Used by CVCFW for corporate service tests. Served as air loads data test aircraft. Delivered in place under bailment to CVCFW as YRB-58A on 14-Dec-58 but incorrectly documented as B-58A. MDS corrected to YRB-58A on 22-Dec-58. Formally delivered to the Air Force (AMC) on 29-Apr-59 but retained at CVCFW. Deployed to Eglin AFB, FL to assist in fire control tests from

Tail #	CN	Lineage	Status	Last Unit	Date	Comments
						Jun-59 to Jan-60. Crashed near Lubbock, TX during flight test on 4-Jun-60 with Convair crew onboard. Enroute from Kirtland AFB, NM to Fort Worth, TX. Loss of control during normal flight. Damaged by hail during a thunderstorm. All three crew members ejected but were killed when open parachutes dragged through rugged terrain in high winds. Total 3 fatalities/0 survivors.
55-0668	9	YRB-58A NB-58A TB-58A	Display	43rd BMW		First flight 13-May-58. Formally accepted by the Air Force on 13-May-58 and delivered in place under bailment to CVCFW as YRB-58A. Incorrectly documented as B-58A. Used by CVCFW for 417.5 hours of corporate service tests including BNS. MDS corrected to YRB-58A on 22-Dec-58. Formally delivered to the Air Force (AMC) on 29-Apr-59 but retained at CVCFW. Redesignated NB-58A on 13-Jan-60 and used for test of Hughes AN/APQ-69 SLAR, Goodyear AN/APS-73 SAR (under Project QUICK CHECK), and other special projects. Fitted with J79-GE-9 engines intended for B-58B. Modified to operational aircraft for SAC, redesignated TB-58A on 13-Apr-64. Last aircraft assigned to 43rd BMW on 14-Apr-64. To MASDC 16-Jan-70. To Southwest Aerospace Museum, Fort Worth, TX. To Meacham Airport for planned new museum facility. Moved to AF Plant 4 when museum plans were dropped. To Lonestar Flight Museum in Galveston, TX in 1990. To Little Rock AFB, AR airpark in 2012. Named "Wild Child II" and "Peeping Tom."
55-0669	10	YRB-58A	Destroyed	CVCFW	27-Oct-59	Delivered in place under bailment to CVCFW as YRB-58A on 22-Apr-58 but incorrectly documented as B-58A. First flight 3-May-58. Formally accepted by the Air Force at CVCFW on 31-May-58. Used by CVCFW for corporate service tests including passive ECM system tests, engine performance tests, and autopilot evaluation tests. MDS corrected to YRB-58A on 22-Dec-58. Formally delivered to the Air Force (AMC) on 29-Apr-59 but retained at CVCFW. Scheduled for conversion to TB-58A but crashed near Hattiesburg, MS about one hour after take-off from Fort Worth, TX on 27-Oct-59 with Convair test crew on board. Loss of control during normal flight. Total 1 fatality/2 survivors.
55-0670	11	YRB-58A TB-58A	Retired	43rd BMW	11-Dec-69	First flight 26-Jun-58. Formally accepted by the Air Force at CVCFW on 26-Jun-58 and delivered in place under bailment to CVCFW as YRB-58A. Incorrectly documented as B-58A.

Tail #	CN	Lineage	Status	Last Unit	Date	Comments
						Used by CVCFW for corporate service tests. To 6592nd TS at Carswell AFB, TX on 3-Jul-58 for R&D testing and incorrectly documented as RB-58A. To Eglin AFB, FL for climatic chamber test from 8-Jul-58 to Sep-58. To CVCFW on 22-Sep-58. Formally delivered to the Air Force (AMC) on 15-Dec-58 but retained at CVCFW. MDS corrected to YRB-58A on 22-Dec-58. To 6592nd TS at Carswell AFB, TX on 31-Dec-58. To CVCFW on 5-Oct-59. TB-58A prototype, first flew on 10-May-60. Modified to operational aircraft for SAC, redesignated TB-58A on 2-Aug-60. Assigned to 43rd BMW on 13-Aug-60. Scrapped 16-Aug-77.
55-0671	12	YRB-58A TB-58A	Retired	43rd BMW	12-Jan-70	Formally accepted by the Air Force on 17-Oct-58 and delivered in place under bailment to CVCFW as YRB-58A. Incorrectly documented as B-58A. First flight 24-Oct-58. Used by CVCFW for corporate service tests. Formally delivered to the Air Force (AMC) on 27-Nov-58 but retained at CVCFW. To 6592nd TS at Carswell AFB, TX on 3-Dec-58 for R&D testing including pod drop and suitability testing. Logged record 179 flying hours during a 58-day test period. MDS corrected to YRB-58A on 22-Dec-58. Scheduled as testbed for original reconnaissance pod before program was cancelled in 1958. Set endurance record on 23-Mar-60 when it flew 11,000 miles in 18 hours and 10 minutes at 620 mph average. Longest lasting single flight ever flown by a B-58. Last aircraft to complete JUNIOR FLASH-UP Phase I on 9-Apr-60. To 43rd BMW at Carswell AFB, TX on 1-Aug-60. To CVCFW on 27-Oct-60. Modified to operational aircraft for SAC, redesignated TB-58A on 30-Jun-61. Assigned to 305th BMW. Later 43rd BMW. Named "Mary Ann" and "All Day All Night." Scrapped on 11-Jul-77.
55-0672	13	YRB-58A TB-58A	Retired	305th BMW	15-Jan-70	Used by CVCFW for corporate service tests. Formally accepted by the Air Force on 29-Sep-58 and delivered in place under bailment to CVCFW as YRB-58A. Incorrectly documented as B-58A. First flight 7-Oct-58. MDS corrected to YRB-58A on 22-Dec-58. Formally delivered to the Air Force (AMC) on 29-Apr-59 but retained at CVCFW. Deployed to Kirtland AFB, NM for pod drop testing. Modified with large datalink antenna on right side of fuselage. Had "Q Twice" on right side and "Clean Sweep" on left side. Modified to operational aircraft for SAC, redesignated TB-58A on 31-Oct-60.

Tail #	CN	Lineage	Status	Last Unit	Date	Comments
						Assigned to 43rd BMW. Later 305th BMW. Named "Lucky 13, "Sweet Sadness", "Q Twice", and "Clean Sweep." Scrapped on 3-Aug-77.
58-1007	14	YRB-58A TB-58A	Retired	43rd BMW	15-Jan-70	Delivered in place under bailment to CVCFW as YRB-58A on 31-Oct-58 but incorrectly documented as B-58A. Used for corporate service tests including BNS development. First flight 8-Nov-58. Formally accepted by the Air Force at CVCFW on 8-Nov-58. MDS corrected to YRB-58A on 22-Dec-58. Formally delivered to the Air Force (AMC) on 29-Apr-59 but retained at CVCFW for R&D testing. To 6592nd TS at Carswell AFB, TX on 30-Jun-60. To CVCFW on 27-Jul-60. Modified to operational aircraft for SAC, redesignated TB-58A on 28-Feb-61. Assigned to 43rd BMW. Named "Super Sue" and "Boomerang." Scrapped on 5-Aug-77.
58-1008	15	YRB-58A	Destroyed	6592nd TS	16-Dec-58	Delivered in place under bailment to CVCFW as YRB-58A on 14-Oct-58 and used for corporate service tests including pod and suitability. Formally accepted by the Air Force at CVCFW on 31-Oct-58. Formally delivered to the Air Force (AMC) on 16-Dec-58 and assigned to 6592nd TS (ARDC) at Carswell AFB, TX for R&D testing. Crashed the same day near Cannon AFB, NM after loss of control due to electrical failure causing inoperative auto-trim and ratio changer. Total 1 fatality/2 survivors. First B-58 lost in an accident.
58-1009	16	YRB-58A B-58A	Retired	43rd BMW	12-Dec-69	Formally accepted by the Air Force on 15-Dec-58 and delivered in place under bailment to CVCFW as YRB-58A for corporate service tests including BNS and production fuel systems. Formally delivered to the Air Force (AMC) on 29-Apr-59 but retained at CVCFW. To 6592nd TS (ARDC) at Carswell AFB, TX on 11-Sep-59 for R&D testing. Had large "16" on fuselage under aft panel of windscreen. To CVCFW on 29-Apr-60. Modified to operational aircraft for SAC under Project JUNIOR FLASH-UP Phase II and redesignated B-58A on 31-Mar-61. Assigned to 43rd BMW. Named "Sweet Sixteen", "El Toro de Moron", and "Bonanza." Scrapped on 9-Jun-77.
58-1010	17	YRB-58A B-58A	Retired	43rd BMW	16-Dec-69	Formally accepted by the Air Force on 10-Jan-59 and delivered in place under bailment to CVCFW as YRB-58A for corporate service tests. Formally delivered to the Air Force (AMC) on 20-Mar-59. To 6592nd TS (ARDC) at Carswell AFB, TX on 21-Mar-59 for R&D

Tail #	CN	Lineage	Status	Last Unit	Date	Comments
						testing. To CVCFW on 29-Feb-60. Modified to operational aircraft for SAC under Project JUNIOR FLASH-UP Phase II and redesignated B-58A on 10-Feb-61. Assigned to 43rd BMW. Named "Hot Stuff." Scrapped on 11-Jul-77.
58-1011	18	YRB-58A B-58A	Retired	43rd BMW	11-Dec-69	Formally accepted by the Air Force on 29-Jan-59 and delivered in place under bailment to CVCFW as YRB-58A for corporate service tests. First flight 30-Jan-59. Formally delivered to the Air Force (AMC) on 29-Apr-59 but retained at CVCFW. First to drop a pod using a fully functional BNS on 12-Feb-60. Modified to operational aircraft for SAC under Project JUNIOR FLASH-UP Phase II and redesignated B-58A on 5-May-61. Assigned to 43rd BMW. First 43rd BMW aircraft to arrive at Little Rock AFB, AR on 1-Sep-64. Named "Wicked Witch", "Trailblazer", "The Pulaski Hustler" and "The Polish Prostitute." Scrapped on 6-Jun-77.
58-1012	19	YRB-58A	Destroyed	CVCFW	14-May-59	Formally accepted by the Air Force on 17-Feb-59 and delivered in place under bailment to CVCFW as YRB-58A for corporate service tests including BNS testing. Formally delivered to the Air Force (AMC) on 4-May-59 but retained at CVCFW. Destroyed by ground fire at CVCFW on 14-May-59 after a fuel leak and accidental ignition while transferring fuel. Total 2 fatalities/13 injured on ground.
58-1013	20	YRB-58A B-58A	Retired	43rd BMW	29-Dec-69	Formally accepted by the Air Force on 23-Feb-59 and delivered in place under bailment to CVCFW as YRB-58A for corporate service tests. Formally delivered to the Air Force (AMC) on 26-Apr-59. To 6592nd TS (ARDC) at Carswell AFB, TX on 27-Apr-59 for R&D testing. First aircraft to enter and complete JUNIOR FLASH-UP Phase I in Feb-60. To CVCFW on 1-May-61. Modified to operational aircraft for SAC under Project JUNIOR FLASH-UP Phase II and redesignated B-58A on 31-Oct-61. Assigned to 43rd BMW. Scrapped on 25-Jul-77.
58-1014	21	YRB-58A B-58A	Retired	43rd BMW	7-Jan-70	Formally accepted by the Air Force on 12-Apr-59 and delivered in place under bailment to CVCFW as YRB-58A for corporate service tests. Formally delivered to the Air Force (AMC) on 29-Jun-59 and assigned to 6592nd TS (ARDC) at Carswell AFB, TX for R&D testing. Completed Project JUNIOR FLASH-UP modifications in Mar-60. To CVCFW on 1-Nov-60. Modified to operational aircraft for SAC under Project JUNIOR FLASH-UP Phase II and

Tail #	CN	Lineage	Status	Last Unit	Date	Comments
						redesignated B-58A on 1-Sep-61. Assigned to 43rd BMW. Scrapped on 29-Jun-77.
58-1015	22	YRB-58A B-58A	Retired	43rd BMW	22-Dec-69	First flight 19-Mar-59. Formally accepted by the Air Force on 19-Mar-59 and delivered in place under bailment to CVCFW as YRB-58A for corporate service tests. Formally delivered to the Air Force (AMC) on 29-Apr-59 but retained at CVCFW. Flew low-level mission from Carswell to Edwards AFB, CA at 500 feet and 703 mph on 18-Sep-59. Flew first sustained Mach 2 flight from Seatle to Carswell at nearly 1,320 mph in 70 minutes on 15-Oct-59. To the AFFTC on 10-Dec-59. Damaged when right main landing gear tires failed during take-off on 13-Apr-60. Flew TC pod performance and stability demonstrations on 10-Apr-61. Documented in error as YB-58A during part of time at Edwards. MDS corrected to YRB-58A on 10-Dec-60. To CVCFW on 3-Jan-62. Last aircraft to complete JUNIOR FLASH-UP Phase II modification to operational aircraft for SAC and redesignated B-58A on 25-Oct-62. Assigned to 43rd BMW. Named "Little Joe" and "Ginger." Scrapped on 9-Aug-77.
58-1016	23	YRB-58A B-58A	Destroyed	43rd BMW	20-May-65	Formally accepted by the Air Force on 19-Apr-59 and delivered in place under bailment to CVCFW as YRB-58A for corporate service tests. Formally delivered to the Air Force (AMC) on 20-Jun-59 and assigned to 6592nd TS at Carswell AFB, TX (ARDC) for R&D testing. To CVCFW on 4-Apr-60. Modified to operational aircraft for SAC under Project JUNIOR FLASH-UP Phase II and redesignated B-58A on 24-Mar-61. Assigned to 43rd BMW. Crashed at Little Rock AFB, AR on 20-May-65 due to hard landing. Total 1 fatality/2 survivors. Navigator/bombardier ejected at low speed and was killed. Named "Champion of Champions."
58-1017	24	YRB-58A	Destroyed	6592nd TS	16-Sep-59	Formally accepted by the Air Force on 30-Apr-59 and delivered in place under bailment to CVCFW as YRB-58A for corporate service tests. Formally delivered to the Air Force (AMC) on 15-Jul-59 and assigned to 6592nd TS (ARDC) at Carswell AFB, TX for R&D testing. Destroyed during aborted take-off at Carswell AFB, TX on 16-Sep-59 due to tire failure. Total 2 fatalities/1 survivor.
58-1018	25	YRB-58A B-58A	Retired	43rd BMW	15-Dec-69	First flight 29-Apr-59. Formally accepted by the Air Force on 29-Apr-59 and delivered in place under bailment to CVCFW as YRB-58A for corporate service tests including ECM testing.

Tail #	CN	Lineage	Status	Last Unit	Date	Comments
						Formally delivered to the Air Force (AMC) on 29-Apr-59 but retained at CVCFW. To 6592nd TS (ARDC) at Carswell AFB, TX on 30-Jun-60 for R&D testing. To CVCFW on 2-Oct-60. Damaged on 19-Sep-61 when left main landing gear failed during takeoff at Edwards AFB, CA and Returned to Service (RTS). Modified to operational aircraft for SAC under Project JUNIOR FLASH-UP Phase II and redesignated B-58A on 7-Nov-61. Assigned to 43rd BMW. Named "Reddy Kilowatt" and "Omega." Scrapped on 24-Jun-77.
58-1019	26	YRB-58A B-58A	Retired	43rd BMW	5-Jan-70	Formally accepted by the Air Force on 31-May-59 and delivered in place under bailment to CVCFW as YRB-58A for corporate service tests. Formally delivered to the Air Force (AMC) on 29-Jun-59 but retained at CVCFW. To 6592nd TS (ARDC) at Carswell AFB, TX on 14-Aug-59 for R&D testing. Completed Project JUNIOR FLASH-UP modifications in Mar-60. To CVCFW on 1-Dec-60. Modified to operational aircraft for SAC under Project JUNIOR FLASH-UP Phase II and redesignated B-58A on 17-Aug-61. Assigned to 43rd BMW. Named "Black Dragon" and "Beech-Nut Kid." Scrapped on 28-Jul-77.
58-1020	27	YRB-58A B-58A	Destroyed	43rd BMW	27-Dec-61	Formally accepted by the Air Force on 29-Jun-59 and delivered in place under bailment to CVCFW as YRB-58A for corporate service tests. Formally delivered to the Air Force (AMC) on 5-Oct-59 and assigned to 6592nd TS (ARDC) at Carswell AFB, TX for R&D testing. Completed Project JUNIOR FLASH-UP modifications in Apr-60. To AFCTE for testing on 19-Sep-60. Documented in error as B-58A during time at Edwards. To CVCFW on 28-Dec-60. MDS corrected to YRB-58A. Modified to operational aircraft for SAC under Project JUNIOR FLASH-UP Phase II and redesignated B-58A on 29-Sep-61. Assigned to 43rd BMW. Crashed during normal flight four miles east of Cole Camp, MO on 27-Dec-61 due to engine flame-out caused by ruptured fuel manifold. Crew ejected safely. Total 0 fatalities/3 survivors.
58-1021	28	YRB-58A B-58A	Retired	43rd BMW	15-Dec-69	Formally accepted by the Air Force on 14-Jul-59 and delivered in place under bailment to CVCFW as YRB-58A for corporate service tests. Had large red "2" on vertical fin. Formally delivered to the Air Force (AMC) on 16-Jul-60 but retained at CVCFW. First aircraft to enter and complete JUNIOR FLASH-UP Phase II

Tail #	CN	Lineage	Status	Last Unit	Date	Comments
						modification to operational aircraft for SAC and redesignated B-58A on 30-Dec-60. Assigned to 43rd BMW. As of Oct-65 had highest B-58 flight hours (1,078). Scrapped on 24-Jun-77.
58-1022	29	YRB-58A	Destroyed	CVCFW		Formally accepted by the Air Force on 8-Jul-59 and delivered in place under bailment to CVCFW as YRB-58A for corporate service tests. Formally delivered to the Air Force (AMC) on 9-Oct-60 but retained at CVCFW. Used for cyclic testing Mar-60 to Jan-61. Later tested to destruction during 5-year fatigue test program. Underwent routine loads test in multiple simulated weather conditions and "flew" equivalent of 10 years without structural flaws.
58-1023	30	YRB-58A	Destroyed	CVCFW	22-Apr-60	First flight 24-Jul-59. Formally accepted by the Air Force on 24-Jul-59 and delivered in place under bailment to CVCFW as YRB-58A for corporate service tests including BNS testing. First test aircraft fully configured with tactical equipment. Scheduled to be modified to operational aircraft for SAC and redesignated to B-58A. However, crashed on 22-Apr-60 before modification in the Great Salt Lake during flight test 29 miles NW of Ogden, UT after take-off from Fort Worth, TX with Convair crew onboard. Loss of control due to air data system failure. Total 2 fatalities (pilot and engineer)/1 survivor (engineer).
59-2428	31	B-58A	Retired	43rd BMW	8-Jan-70	First of 86 production B-58A models. First flew Sep-59. First aircraft with operational tail gun. Formally accepted and delivered to the Air Force on 30-Nov-59. Initially assigned to 6592nd TS at Carswell AFB, TX on 1-Dec-59. Flew 1,680 miles in 80 minutes and set a new speed record at an average speed of 1,213.16 mph on 15-Dec-59. Deployed to Ellsworth AFB, SD under Project WHITE HORSE for "cold soak" tests in Jan-60. Assigned to 43rd BMW (SAC) on 1-Aug-60. First aircraft to complete Project HUSTLE-UP Phase I modifications and returned to 43rd BMW on 18-Jan-63. First aircraft to enter Project HUSTLE-UP Phase II on 14-Aug-63. Scrapped on 6-Jul-77.
59-2429	32	B-58A	Retired	43rd BMW	18-Dec-69	Formally accepted by the Air Force on 24-Feb-60. Formally delivered to the Air Force on 26-Feb-60 and initially assigned to 6592nd TS at Carswell AFB, TX. Assigned to 43rd BMW (SAC) on 1-Aug-60. First aircraft to enter SENIOR FLASH-UP Cycle II modification program in Apr-61. Primary aircraft for 1960 SAC Bomb Comp. Second aircraft named "The Pulaski

Tail #	CN	Lineage	Status	Last Unit	Date	Comments
						Hustler" along with 58-1011. Scrapped on 2-Aug-77.
59-2430	33	B-58A	Retired	43rd BMW	13-Jan-70	Formally accepted by the Air Force on 29-Jan-60. Formally delivered to the Air Force on 6-Feb-60 and initially assigned to 6592nd TS at Carswell AFB, TX. Assigned to 43rd BMW (SAC) on 1-Aug-60. Back-up aircraft for 1960 SAC Bomb Comp. Scrapped on 10-Aug-77.
59-2431	34	B-58A	Retired	43rd BMW	8-Dec-69	Formally accepted by the Air Force on 29-Feb-60. Formally delivered to the Air Force on 7-Mar-60 and initially assigned to 6592nd TS at Carswell AFB, TX. Assigned to 43rd BMW (SAC) on 1-Aug-60. Flew at Mach 2.0 for 78 minutes in Oct-60. Longest Mach 2.0 flight at that time. Scrapped on 29-Jun-77.
59-2432	35	B-58A	Retired	43rd BMW	19-Dec-69	Formally accepted by the Air Force on 24-Mar-60. Formally delivered to the Air Force on 28-Mar-60 and initially assigned to 6592nd TS at Carswell AFB, TX. Assigned to 43rd BMW (SAC) on 1-Aug-60. Named "Regal Beagle." Scrapped on 1-Aug-77.
59-2433	36	B-58A	Retired	43rd BMW	7-Jan-70	Formally accepted by the Air Force on 29-Apr-60. Formally delivered to the Air Force on 5-May-60 and initially assigned to 6592nd TS at Carswell AFB, TX. Assigned to 43rd BMW (SAC) on 1-Aug-60. Named "Now or Never." Scrapped on 20-Jul-77.
59-2434	37	B-58A	Retired	43rd BMW	17-Dec-69	Formally accepted by the Air Force on 13-Apr-60. Formally delivered to the Air Force on 14-Apr-60 and initially assigned to 6592nd TS at Carswell AFB, TX. Assigned to 43rd BMW (SAC) on 1-Aug-60. First to enter SENIOR FLASH-UP Cycle I on 2-Sep-60. Named "Cannonball." Scrapped on 24-Jun-77.
59-2435	38	B-58A	Retired	43rd BMW	7-Jan-70	Formally accepted by the Air Force on 31-May-60 and delivered in place to CVCFW. Used for TC pod and multi-weapons drop testing at Kirtland AFB, NM and CVCFW. Had 17 small component pod symbols, 11 big, and 40 bomb symbols painted on nose. Made first upper component drop at White Sands Missile Range in Oct-60. First Mach 2.0 upper component drop on 10-Feb-61. Dropped lower component on first pass over White Sands Missile Range, NM and upper component on second pass during test in May-61. Flew multi-weapon aerodynamic and ballistic reaction tests in Jun-61. Flew last Category III test flight on 31-Jul-61. First Mach 2.0 lower pod drop on 8-Aug-61. Successful supersonic pod drop over water at Mach 1.6 and 35,000 feet near Eglin AFB, FL

Tail #	CN	Lineage	Status	Last Unit	Date	Comments
						in Sep-61. First multi-weapons drop on 7-Feb-62 over White Sands. First supersonic multi-weapons drop on 19-Jun-62. First Mach 2.0 multi-weapons drop on 2-Aug-62. Completed pod drop testing at Kirtland with two high altitude TC pod drops in Mar-63 followed by two more drops over Eglin Gulf Test Range. Assigned to 43rd BMW (SAC) on 1-Nov-63. Named "Shackbuster." Scrapped 27-Jun-77.
59-2436	39	B-58A	Retired	43rd BMW	9-Jan-70	First fully operational B-58A equipped with all tactical systems. Formally accepted by the Air Force on 31-Aug-60. Formally delivered to the Air Force on 3-Sep-60 and assigned to 43rd BMW (SAC). Scrapped on 5-Jul-77.
59-2437	40	B-58A GB-58A	Display	43rd BMW	16-Jul-68	Formally accepted by the Air Force on 26-Oct-60. Formally delivered to the Air Force on 31-Oct-60 and assigned to 43rd BMW (SAC). Damaged beyond repair during landing at Little Rock AFB, AR on 16-Jul-68. Permanently grounded after main landing gear strut failure and used as a training aid. Total 0 fatalities/3 survivors. Possibly redesignated GB-58A (no entry in aircraft record cards). Put on display at Kelly AFB, TX main gate (now Port San Antonio). Named "Firefly II" and Rigley's Baby."
59-2438	41	B-58A	Retired	43rd BMW	19-Dec-69	Formally accepted by the Air Force on 30-Sep-60. Formally delivered to the Air Force on 3-Oct-60 and assigned to 43rd BMW (SAC). Sometimes carried LA-1 reconnaissance pod. Scrapped on 25-May-77.
59-2439	42	B-58A	Retired	43rd BMW	12-Jan-70	Formally accepted by the Air Force on 25-Oct-60 and delivered in place under bailment to CVCFW. Used for multi-weapons drop testing at Kirtland AFB, NM and captive flights from CVCFW to determine performance characteristics of the new configuration. Flew multi-weapon aerodynamic and ballistic reaction tests in Jun-61. Used for ground suspended vibration tests and flutter analysis in Nov-61. Conducted multi-weapons Phase II flight demonstrations in Jan-62. Assigned to 43rd BMW (SAC) on 5-Nov-63. Scrapped 1-Aug-77.
59-2440	43	B-58A	Retired	43rd BMW	8-Jan-70	Formally accepted by the Air Force on 28-Oct-60. Formally delivered to the Air Force on 31-Oct-60 and assigned to 43rd BMW (SAC). Possibly last to visit UK in May-69. Scrapped on 6-Jul-77.
59-2441	44	B-58A	Retired	43rd BMW	5-Jan-70	Formally accepted by the Air Force on 1-Nov-60. Formally delivered to the Air Force on 8-Nov-60 and assigned to 43rd BMW (SAC). Broke three speed records on 14-Jan-61 during

Tail #	CN	Lineage	Status	Last Unit	Date	Comments
						Operation QUICK STEP II that were previously set by B-58A (59-2442) on 12-Jan-61. Awarded Thompson Trophy. Named "Roadrunner." Scrapped on 4-Aug-77.
59-2442	45	B-58A	Retired	43rd BMW	12-Nov-69	Formally delivered to the Air Force on 12-Dec-60 and assigned to 43rd BMW (SAC). Set six new speed with payload records on 12-Jan-61 during Operation QUICK STEP I. Crew awarded Distinguished Flying Cross. Named "Untouchable." Scrapped on 9-Aug-77.
59-2443	46	B-58A	Destroyed	43rd BMW	15-Jun-65	Formally accepted by the Air Force on 9-Jan-61. Formally delivered to the Air Force on 13-Jan-61 and assigned to 43rd BMW (SAC). Crashed on 15-Jun-65 during landing for Paris Air Show at Le Bourget Airport, France. Overweight due to excess fuel still onboard, crashed short of the runway, hit instrument approach beacons, and burst into flames. Pilot killed. Named "Bye Bye Birdie." Total 1 fatality/2 survivors.
59-2444	47	B-58A	Retired	43rd BMW	10-Dec-69	Formally accepted by the Air Force on 31-Dec-60. Formally delivered to the Air Force on 5-Jan-61 and assigned to 43rd BMW (SAC). Damaged during landing at Little Rock AFB, AR on 20-May-65 when nose gear failed to extend. Jettisoned pod prior to landing. RTS at Little Rock. Named "Lucky Lady V." Scrapped on 27-May-77.
59-2445	48	B-58A	Retired	43rd BMW	17-Dec-69	Formally accepted by the Air Force on 23-Jan-61. Formally delivered to the Air Force on 27-Jan-61 and assigned to 43rd BMW (SAC). Named "Sno White." Scrapped on 28-Jun-77.
59-2446	49	B-58A	Retired	43rd BMW	3-Nov-69	Formally accepted by the Air Force on 20-Feb-61. Formally delivered to the Air Force on 21-Feb-61 and assigned to 43rd BMW (SAC). First aircraft retired. Scrapped on 19-Jul-77.
59-2447	50	B-58A	Destroyed	43rd BMW	15-Feb-62	Formally accepted by the Air Force on 9-Dec-60. Formally delivered to the Air Force on 12-Dec-60 and assigned to 43rd BMW (SAC). Crashed on 15-Feb-62 during routine training flight near Lawton, OK after loss of control due to Mach and airspeed system malfunction. Aircraft began losing altitude at 35,000 ft. Pilot ordered crew to eject at 15,000 ft. All crew ejected safely. Total 0 fatalities/3 survivors. Named "Rapid Rabbit."
59-2448	51	B-58A	Retired	43rd BMW	20-Nov-69	Formally accepted by the Air Force on 16-Mar-61. Formally delivered to the Air Force on 21-Mar-61 and assigned to 43rd BMW (SAC). Scrapped on 27-May-77.

Tail #	CN	Lineage	Status	Last Unit	Date	Comments
59-2449	52	B-58A	Retired	43rd BMW	13-Jan-70	Formally accepted by the Air Force on 27-Feb-61. Formally delivered to the Air Force on 6-Mar-61 and assigned to 43rd BMW (SAC). Damaged during landing in inclement weather in Oct-62. Repaired at GDFW and RTS. First aircraft entered into Project HUSTLE-UP Phase I on 30-Oct-62. Named "Hobo 49." Scrapped on 12-Jul-77.
59-2450	53	B-58A	Retired	43rd BMW	15-Jan-70	Formally accepted by the Air Force on 25-Jan-61. Formally delivered to the Air Force on 26-Jan-61 and assigned to 43rd BMW (SAC). Scrapped on 16-Jun-77.
59-2451	54	B-58A	Destroyed	43rd BMW	3-Jun-61	Formally accepted by the Air Force on 27-Jan-61. Formally delivered to the Air Force on 31-Jan-61 and assigned to 43rd BMW (SAC). Crashed on 3-Jun-61 while performing low-level aerobatic flight during the Paris Airshow at Le Bourget Airport, France. Total 3 fatalities/0 survivors. The aircraft, named "The Firefly" had just completed the first supersonic transatlantic crossing from New York to Paris on 26-May-61 in less than 3 hours and 15 minutes. Also set record for sustained speed on 10-May-61 flying 669.4 miles in 30 minutes and 43 seconds at an average speed of 1,302.07 mph (2,095 kmph).
59-2452	55	B-58A	Retired	43rd BMW	9-Jan-70	Formally accepted by the Air Force on 22-Mar-61. Formally delivered to the Air Force on 22-Mar-61 and assigned to 43rd BMW (SAC). Scrapped on 27-May-77.
59-2453	56	B-58A	Retired	43rd BMW	5-Jan-70	Formally accepted by the Air Force on 11-Mar-61. Formally delivered to the Air Force on 14-Mar-61 and assigned to 43rd BMW (SAC). Named "Top Cat." Scrapped on 22-Aug-77.
59-2454	57	B-58A	Retired	43rd BMW	14-Jan-70	Formally accepted by the Air Force on 31-May-61. Formally delivered to the Air Force on 1-Jun-61 and assigned to 43rd BMW (SAC). Achieved 31 consecutive on-time takeoffs in 1962 demonstrating improving B-58A maintenance performance and aircraft reliability. Damaged during taxi at Little Rock AFB, AR due to structural failure of the forward fuselage and RTS. Named "Wild Child" and "Patches." Scrapped on 23-Aug-77.
59-2455	58	B-58A	Retired	43rd BMW	15-Jan-70	Formally accepted by the Air Force on 20-Apr-61. Formally delivered to the Air Force on 24-Apr-61 and assigned to 43rd BMW (SAC). Scrapped on 30-Jun-77.
59-2456	59	B-58A	Retired	43rd BMW	9-Dec-69	Formally accepted by the Air Force on 30-Jan-61 and delivered in place CVCFW. Assigned to the AFFTC on 20-Sep-61. Had "Q" on left side

Tail #	CN	Lineage	Status	Last Unit	Date	Comments
						of nose. Conducted sonic-boom testing from Sep-61 to Mar-62. Conducted engine out take-off and landing tests. Conducted preliminary multi-weapons testing. Set two altitude records on 14-Sep-62 carrying a payload of 2,000 kg and 5,000 kg to 85,360.66 feet and won the 1962 Harmon Trophy. Final Project HUSTLE-UP Phase I aircraft completed in Aug-63. Assigned to 43rd BMW (SAC) on 22-Nov-63. Scrapped on 1-Jun-77.
59-2457	60	B-58A	Retired	43rd BMW	8-Jan-70	Formally accepted by the Air Force on 19-Apr-61. Formally delivered to the Air Force on 20-Apr-61 and assigned to 43rd BMW (SAC). Scrapped on 27-Jun-77.
59-2458	61	B-58A	Display	305th BMW		Formally accepted by the Air Force on 29-Apr-61. Formally delivered to the Air Force on 3-May-61 and assigned to 43rd BMW (SAC). Set three world speed records on 5-Mar-62 during Operation HEAT RISE for round trip flight between Los Angeles and New York City and received Mackay Trophy. Also received the Bendix Trophy for the Los Angeles to New York City outbound flight that same day. Later reassigned to 305th BMW. Currently on display at the National Museum of the Unted States Air Force. Named "Cowtown Hustler."
59-2459	62	B-58A	Destroyed	43rd BMW	5-Mar-62	Formally accepted by the Air Force on 29-Mar-61. Formally delivered to the Air Force on 31-Mar-61 and assigned to 43rd BMW (SAC). Second aircraft for Operation HEAT RISE round trip flight from Los Angeles to New York on 5-Mar-62. Crashed on take-off from Carswell AFB, TX later that day due to mechanical failure of the flight control system. Reports said the aircraft lost one wing and went into a 6,000-foot skid that left the aircraft upside down and burning on an ammunition dump. One crew member ejected but was not thrown clear of the wreck and fire. Total 3 fatalities/0 survivors.
59-2460	63	B-58A	Retired	43rd BMW	18-Nov-69	Formally accepted by the Air Force on 29-Apr-61. Formally delivered to the Air Force on 4-May-61 and assigned to 43rd BMW (SAC). Displayed for President Kennedy during firepower display at Eglin AFB, FL. Scrapped 6-Jun-77.
59-2461	64	B-58A	Retired	43rd BMW	14-Jan-70	Formally accepted by the Air Force on 9-May-61. Formally delivered to the Air Force on 11-May-61 and became first B-58A assigned to 305th BMW. Later reassigned to 43rd BMW. Named "Hoosier Hustler." Temporarily painted to look like "The Firefly" (which had been destroyed) for a film. Scrapped on 19-Aug-77.

Tail #	CN	Lineage	Status	Last Unit	Date	Comments
59-2462	65	B-58A	Destroyed	305th BMW	12-Apr-62	Formally accepted by the Air Force on 27-Jul-61. Formally delivered to the Air Force on 2-Aug-61 and assigned to 305th BMW (SAC). Crashed on 12-Apr-62 shortly after take-off from Bunker Hill AFB, IN due to flight control system failure. Two crew ejected safely. Total 1 fatality (navigator/bombardier)/2 survivors. Resulted in modifications to remaining fleet.
59-2463	66	B-58A	Retired	43rd BMW	7-Nov-69	Formally accepted by the Air Force on 9-May-61. Formally delivered to the Air Force on 12-May-61 and assigned to 43rd BMW (SAC). Last aircraft to complete Project SENIOR FLASH-UP Cycle II on 9-Aug-62. Reassigned to 305th BMW and later returned to 43rd BMW. Named "The Heart of Dixie." Scrapped on 10-Jun-77.
60-1110	67	B-58A	Retired	305th BMW	18-Dec-69	Formally accepted by the Air Force on 8-Jun-61. Formally delivered to the Air Force on 14-Jun-61 and assigned to 305th BMW (SAC). Originally scheduled as production prototype for the B-58B with four J79-GE-9 engines but the program was cancelled. Built as B-58A instead. Named "City of Peru." Scrapped on 18-Aug-77.
60-1111	68	B-58A	Retired	305th BMW	8-Dec-69	Formally accepted by the Air Force on 8-Jun-61. Formally delivered to the Air Force on 12-Jun-61 and assigned to 305th BMW (SAC). Later reassigned to 43rd BMW and then returned to 305th BMW. Named "Four Aces." Scrapped on 8-Jun-77.
60-1112	69	B-58A	Retired	305th BMW	5-Jan-70	Formally accepted by the Air Force on 27-Jul-61. Formally delivered to the Air Force on 1-Aug-61 and assigned to 305th BMW (SAC). Later reassigned to 43rd BMW and then returned to 305th BMW. Scrapped on 15-Jul-77.
60-1113	70	B-58A	Retired	305th BMW	16-Dec-69	Formally accepted by the Air Force on 28-Jun-61. Formally delivered to the Air Force on 30-Jun-61 and assigned to 305th BMW (SAC). Later reassigned to 43rd BMW and then returned to 305th BMW. Scrapped on 28-Jun-77.
60-1114	71	B-58A	Retired	305th BMW	17-Nov-69	Formally accepted by the Air Force on 31-Jul-61. Formally delivered to the Air Force on 8-Aug-61 and assigned to 305th BMW (SAC). Later reassigned to 43rd BMW and then returned to 305th BMW. Last HUSTLE-UP Phase II aircraft completed on 28-Dec-63. Had "State of Texas" painted on nose during flight test at CVCFW. Scrapped on 8-Jun-77.
60-1115	72	B-58A	Retired	305th BMW	13-Jan-70	Formally accepted by the Air Force on 28-Aug-61. Formally delivered to the Air Force on 31-Aug-61 and assigned to 305th BMW (SAC). Later reassigned to 43rd BMW and then

Tail #	CN	Lineage	Status	Last Unit	Date	Comments
						returned to 305th BMW. Scrapped on 21-Jul-77.
60-1116	73	B-58A	Destroyed	305th BMW	8-Dec-64	Formally accepted by the Air Force on 27-Jul-61. Formally delivered to the Air Force on 2-Aug-61 and assigned to 305th BMW (SAC). Destroyed on 8-Dec-64 due to collapsed landing gear and fire after the aircraft slid off icy taxiway during training exercise at Bunker Hill AFB, IN. One crew member ejected but was killed. Total 1 fatality (nav/bombardier)/2 survivors. Five nuclear weapons on board were burned and resulted in radioactive contamination of the crash area. Some portions of the weapons were melted and burned but did not detonate. One weapon was extinguished by placing it in a shallow trench and covering it with sand. All radioactive contamination was removed and buried with the aircraft on a remote location of the base.
60-1117	74	B-58A	Retired	305th BMW	14-Jan-70	Formally accepted by the Air Force on 23-Aug-61. Formally delivered to the Air Force on 30-Aug-61 and assigned to 305th BMW (SAC). Scrapped on 2-Aug-77.
60-1118	75	B-58A	Retired	305th BMW	15-Dec-69	Formally accepted by the Air Force on 15-Sep-61. Formally delivered to the Air Force on 15-Sep-61 and assigned to 305th BMW (SAC). Scrapped on 27-Jun-77.
60-1119	76	B-58A	Destroyed	305th BMW	12-Dec-66	Formally accepted by the Air Force on 18-Aug-61. Formally delivered to the Air Force on 23-Aug-61 and assigned to 305th BMW (SAC). Modified to carry iron bombs during Operation BULLSEYE in Apr-64. Crashed into the ground on 12-Dec-66 during low-level bombing run 70 km SSW of Lexington, KY (near McKinney, KY). Total 3 fatalities/0 survivors. Named "Pink Panther" and "City of Kokomo."
60-1120	77	B-58A	Retired	305th BMW	12-Dec-69	Formally accepted by the Air Force on 21-Sep-61. Formally delivered to the Air Force on 28-Sep-61 and assigned to 305th BMW (SAC). Scrapped on 14-Jun-77.
60-1121	78	B-58A	Retired	305th BMW	22-Dec-69	Formally accepted by the Air Force on 31-Oct-61. Formally delivered to the Air Force on 20-Nov-61 and assigned to 305th BMW (SAC). Named "Can Do." Scrapped on 11-Aug-77.
60-1122	79	B-58A	Retired	305th BMW	19-Nov-69	Formally accepted by the Air Force on 17-Nov-61. Formally delivered to the Air Force on 20-Nov-61 and assigned to 305th BMW (SAC). Scrapped on 3-Jun-77.
60-1123	80	B-58A	Retired	305th BMW	5-Jan-70	Formally accepted by the Air Force on 29-Nov-61. Formally delivered to the Air Force on 1-Dec-61 and assigned to 305th BMW (SAC). Scrapped on 4-Aug-77.

Tail #	CN	Lineage	Status	Last Unit	Date	Comments
60-1124	81	B-58A	Retired	305th BMW	11-Dec-69	Formally accepted by the Air Force on 21-Sep-61. Formally delivered to the Air Force on 26-Sep-61 and assigned to 305th BMW (SAC). Scrapped on 14-Jun-77.
60-1125	82	B-58A	Retired	305th BMW	8-Dec-69	Formally accepted by the Air Force on 22-Dec-61. Formally delivered to the Air Force on 28-Dec-61 but retained at CVCFW. Assigned to 305th BMW (SAC) on 25-Feb-62. Scrapped on 3-Jun-77.
60-1126	83	B-58A	Retired	305th BMW	7-Jan-70	Formally accepted by the Air Force on 31-Oct-61. Formally delivered to the Air Force on 31-Oct-61 and assigned to 305th BMW (SAC). Scrapped on 13-Jul-77.
60-1127	84	B-58A	Retired	305th BMW	18-Dec-69	Formally accepted by the Air Force on 27-Dec-61. Formally delivered to the Air Force on 29-Dec-61 and assigned to 305th BMW (SAC) on 9-Jan-62. Scrapped on 12-Aug-77.
60-1128	85	B-58A	Destroyed	305th BMW	22-Jul-65	Formally accepted by the Air Force on 27-Oct-61. Formally delivered to the Air Force on 31-Oct-61 and assigned to 305th BMW (SAC). Crashed on 22-Jul-65 during landing at Bunker Hill AFB, IN due to aircraft bouncing and leaving the runway during landing. Total 0 fatalities/3 survivors.
60-1129	86	B-58A	Retired	305th BMW	9-Jan-70	Formally accepted by the Air Force on 29-Nov-61. Formally delivered to the Air Force on 1-Dec-61 and assigned to 305th BMW (SAC). Scrapped on 15-Aug-77.
61-2051	87	B-58A	Retired	43rd BMW	14-Nov-69	Formally accepted by the Air Force on 1-Dec-61. Formally delivered to the Air Force on 4-Dec-61 and assigned to 305th BMW (SAC). Later assigned to 43rd BMW (SAC). Scrapped on 13-Jun-77.
61-2052	88	B-58A	Retired	305th BMW	13-Jan-70	Formally accepted by the Air Force on 28-Dec-61. Formally delivered to the Air Force on 3-Jan-62 and assigned to 305th BMW (SAC). Scrapped on 8-Aug-77.
61-2053	89	B-58A	Retired	305th BMW	6-Jan-70	Formally accepted by the Air Force on 29-Mar-62. Formally delivered to the Air Force on 9-Apr-62 and assigned to 305th BMW (SAC). Scrapped on 28-Jul-77.
61-2054	90	B-58A	Retired	305th BMW	13-Nov-69	Formally accepted by the Air Force on 5-Jan-62. Formally delivered to the Air Force on 10-Jan-62 and assigned to 305th BMW (SAC). Scrapped on 9-Jun-77.
61-2055	91	B-58A	Retired	305th BMW	22-Dec-69	Formally accepted by the Air Force on 20-Jan-62. Formally delivered to the Air Force on 20-Jan-62 and assigned to 305th BMW (SAC). Scrapped on 22-Jul-77.
61-2056	92	B-58A	Destroyed	305th BMW	18-Apr-69	Formally accepted by the Air Force on 12-Mar-62. Formally delivered to the Air Force on 15-

Tail #	CN	Lineage	Status	Last Unit	Date	Comments
						Mar-62 and assigned to 305th BMW (SAC) on 16-Mar-62. Crashed on 18-Apr-69 during normal flight near Danville, IL due to suspected system anomalies. Crew ejected safely. Total 0 fatalities/3 survivors.
61-2057	93	B-58A	Destroyed	305th BMW	14-Sep-62	Formally accepted by the Air Force on 22-Jan-62. Formally delivered to the Air Force on 24-Jan-62 and assigned to 305th BMW (SAC). Crashed on 14-Sep-62 during normal flight three miles east of Butlerville, IN due to structural breakdown caused by flight control system failure. Total 3 fatalities/0 survivors. Last B-58 lost in an accident.
61-2058	94	B-58A	Retired	305th BMW	21-Nov-69	Formally accepted by the Air Force on 15-Feb-62. Formally delivered to the Air Force on 20-Feb-62 and assigned to 305th BMW (SAC). Scrapped on 1-Jun-77.
61-2059	95	B-58A	Display	305th BMW		Formally accepted by the Air Force on 16-Feb-62. Formally delivered to the Air Force on 19-Feb-62 and assigned to 305th BMW (SAC) on 25-Mar-62. Set world speed record on 16-Oct-63 for flight from Tokyo to London and dubbed "Greased Lightning." Previously named "Can Do." Currently on display at the Strategic Air Command and Aerospace Museum.
61-2060	96	B-58A	Retired	305th BMW	17-Dec-69	Formally accepted by the Air Force on 12-Mar-62. Formally delivered to the Air Force on 15-Mar-62 and assigned to 305th BMW (SAC). Scrapped on 2-Aug-77.
61-2061	97	B-58A	Destroyed	305th BMW	14-Jun-67	Formally accepted by the Air Force on 21-Jun-62. Formally delivered to the Air Force on 3-Aug-62 and assigned to 305th BMW (SAC). Crashed at night on 14-Jun-67 west of Darrouzett, TX after the crew ejected due to minor weather damage. Windscreen shattered blinding the pilot who could fly the aircraft but was unable to see for landing. One crew member killed when chute failed to open. Total 1 fatality/2 survivors.
61-2062	98	B-58A	Destroyed	305th BMW	18-Apr-68	First production B-58A to fly with crew capsules installed on 2-Mar-62. Formally accepted by the Air Force on 30-Apr-62. Formally delivered to the Air Force on 17-May-62 and assigned to 305th BMW (SAC). Crashed on 18-Apr-68 during initial take-off climb from Bunker Hill AFB, IN after loss of control due to mechanical failure. Total 3 fatalities/0 survivors.
61-2063	99	B-58A	Destroyed	305th BMW	26-Aug-63	Formally accepted by the Air Force on 14-Aug-62. Formally delivered to the Air Force on 15-Aug-62 and assigned to 305th BMW (SAC). Crashed on 26-Aug-63 during landing at

Tail #	CN	Lineage	Status	Last Unit	Date	Comments
						Bunker Hill AFB, IN due to hard landing. The plane bounced as the drag chute was deployed and fire resulted. Total 2 fatalities/1 survivor. The two crewmembers killed (nav and DSO) when they tried to escape over the wing while the pilot survived by exiting over the nose.
61-2064	100	B-58A	Retired	305th BMW	14-Jan-70	Formally accepted by the Air Force on 30-Apr-62. Formally delivered to the Air Force on 17-May-62 and assigned to 305th BMW (SAC). Scrapped on 5-Jul-77.
61-2065	101	B-58A	Destroyed	305th BMW	13-Nov-67	Formally accepted by the Air Force on 27-Aug-62. Formally delivered to the Air Force on 29-Aug -62 and assigned to 305th BMW (SAC) on 31-Aug-62. Crashed on 13-Nov-67 during take-off from Bunker Hill AFB, IN due to loss of control in initial climb. Total 3 fatalities/0 survivors.
61-2066	102	B-58A JB-58A B-58A	Retired	305th BMW	7-Jan-70	Formally delivered to the Air Force (Systems Command) as B-58A on 31-May-62 but retained at GDFW. To the AFFTC on 29-Nov-62 for testing. Redesignated JB-58A on 7-Oct-63. Fitted with experimental pneumatic seats designed to improve crew comfort. Assigned to 305th BMW as SAC operational aircraft on 12-Aug-64. Scrapped on 13-Jun-77.
61-2067	103	B-58A	Retired	305th BMW	9-Jan-70	Formally accepted by the Air Force on 27-Aug-62. Formally delivered to the Air Force on 7-Dec-62 and assigned to 305th BMW (SAC). Scrapped on 30-Jun-77.
61-2068	104	B-58A	Retired	305th BMW	10-Dec-69	Formally accepted by the Air Force on 31-Aug-62. Formally delivered to the Air Force on 7-Sep-62 and assigned to 305th BMW (SAC). Named "Deputy Dog." Scrapped on 9-Jun-77.
61-2069	105	B-58A	Retired	305th BMW	9-Dec-69	Formally accepted by the Air Force on 29-Aug-62. Formally delivered to the Air Force on 31-Aug-62 and assigned to 305th BMW (SAC). Scrapped on 24-May-77.
61-2070	106	B-58A	Retired	305th BMW	12-Jan-70	Formally accepted by the Air Force on 21-Sep-62. Formally delivered to the Air Force on 21-Sep-62 and assigned to 305th BMW (SAC) on 4-Dec-62. Scrapped on 17-Aug-77.
61-2071	107	B-58A	Retired	305th BMW	15-Jan-70	Formally accepted by the Air Force on 31-Aug-62. Formally delivered to the Air Force on 24-Oct-62 and assigned to 305th BMW (SAC). Scrapped on 19-Jul-77.
61-2072	108	B-58A	Retired	305th BMW	12-Dec-69	Formally accepted by the Air Force on 31-Aug-62. Formally delivered to the Air Force on 5-Sep-62 and assigned to 305th BMW (SAC). Scrapped on 26-Jul-77.
61-2073	109	B-58A	Destroyed	305th BMW	3-Apr-69	Formally accepted by the Air Force on 9-Oct-62. Formally delivered to the Air Force on 12-Oct-62 and assigned to 305th BMW (SAC) on

Tail #	CN	Lineage	Status	Last Unit	Date	Comments
						4-Dec-62. Crashed on 3-Apr-69 during normal flight 15 km south of Lincoln AFB, NE (near Rokeby) due to system failure. Front latch of weapons pod broke causing nose of pod to drop and aircraft nose to pitch up. Left wing broke off when the aircraft was at 45 degrees pitch up and landed some distance from the main wreckage. Crew ejected safely. Total 0 fatalities/3 survivors.
61-2074	110	B-58A	Retired	305th BMW	24-Nov-69	Formally accepted by the Air Force on 28-Sep-62. Formally delivered to the Air Force on 28-Sep-62 and assigned to 305th BMW (SAC) on 4-Dec-62. Scrapped on 23-May-77.
61-2075	111	B-58A	Retired	305th BMW	12-Jan-70	Formally accepted by the Air Force on 4-Oct-62. Formally delivered to the Air Force on 10-Oct-62 and assigned to 305th BMW (SAC). Scrapped on 15-Jul-77.
61-2076	112	B-58A	Retired	305th BMW	13-Jan-70	Formally accepted by the Air Force on 28-Sep-62. Formally delivered to the Air Force on 4-Oct-62 and assigned to 305th BMW (SAC). Scrapped on 16-Jun-77.
61-2077	113	B-58A	Retired	305th BMW	10-Dec-69	Formally accepted by the Air Force on 15-Oct-62. Formally delivered to the Air Force on 15-Oct-62 and assigned to 305th BMW (SAC) on 26-Nov-62. Scrapped on 26-Jul-77.
61-2078	114	B-58A	Retired	305th BMW	16-Jan-70	Formally accepted by the Air Force on 25-Oct-62. Formally delivered to the Air Force on 26-Oct-62 and assigned to 305th BMW (SAC). Named "Top Dawg." Scrapped on 5-Aug-77.
61-2079	115	B-58A	Retired	305th BMW	19-Dec-69	Formally accepted by the Air Force on 25-Oct-62. Formally delivered to the Air Force on 26-Oct-62 and assigned to 305th BMW (SAC). Named "The Thumper." Scrapped 25-May-77.
61-2080	116	B-58A	Display	305th BMW	6-Jan-70	Last B-58A built. First flight 23-Oct-62. Formally accepted by the Air Force on 26-Oct-62. Formally delivered to the Air Force on 26-Oct-62 and assigned to 305th BMW (SAC). To MASDC on 6-Jan-70. To Pima Air & Space Museum.

Bibliography

Brewer, A. P. (1994-2024). *The B-58 Hustler Page*. Retrieved from B-58.com: https://b-58.com/history_origin.php

Griswold, W. S. (1961, August). World's Fastest Bomber. *Popular Science*.

Hehs, E. (2011, July 16). Super Hustler, FISH, Kingfish, and Beyond (Part 2: FISH). *Code One Magazine*.

Hirsch, H. R. (1964, Sept-Oct). The B-58. *Air University Review.*

Holt, G. J. (2023). *The B-58 Blunder, 2nd Edition.*

Knaack, M. S. (1988). *Post-World War II Bombers 1945-1973.* Washington DC: Office of Air Force History.

Miller, J. (1985). *Convair B-58.* Arlington, Texas: Aerofax.

Peacock, L. (1987). *B-52 Stratofortress.* London: Osprey Publishing Ltd.

Robinson, D. H. (1967). *The B-58 Hustler .* New York: Arco Publishing Company, Inc.

Smith, D. (1964, Sept-Oct). B-58 Wing Operations. *Air University Review.*

Notes and Citations

Design and Development

[1] (Brewer, 1994-2024)

[2] Convair became a division of its new parent company General Dynamics during a merger that completed on 29 April 1954. It continued using the Convair product name until September 1961 when it began using General Dynamics for all of its new products such as the F-111 and F-16.

[3] (Knaack, 1988, pp. 351, 353)

[4] (Knaack, 1988, pp. 352, 362)

[5] (Knaack, 1988, p. 351)

[6] (Knaack, 1988, pp. 354, 357)

[7] (Knaack, 1988, pp. 356-357)

[8] (Miller, 1985, p. 19)

[9] (Robinson, 1967)

[10] (Miller, 1985, p. 20)

[11] (Robinson, 1967)

[12] (Knaack, 1988, p. 361)

[13] (Knaack, 1988, pp. 357-358)

[14] (Knaack, 1988, p. 359)

[15] The Air Force subsequently issued a separate directive for a high-subsonic low-level bomber. This was competed separately. A contract was awarded to Martin for a proposed delta-wing aircraft, but it was cancelled in 1957. SAC believed the B-47 and modified B-52 could perform the low-level mission.

[16] (Knaack, 1988, p. 363)

[17] (Brewer, 1994-2024)

[18] (Miller, 1985, p. 26)

[19] If the schedule held the first B-58 wing with 45 aircraft assigned would be operational by early 1959.

[20] Developed in the late 1940s by Major Generals Orval R. Cook, DCS Materiel and Laurance C. Craigie, DCS Development. Both knew it could be expensive and should only be used when there is a high confidence of production.

[21] (Knaack, 1988, pp. 266-267)

[22] The F-102 design was finalized in mid-1956 and resulted in the "coke bottle" or "wasp waist" configuration to equalize cross section areas. The Transonic Area Rule became an accepted facet of aerodynamic design by the time the B-58 prototype first flew on 11 Nov 56.

[23] (Knaack, 1988, p. 367)

[24] (Knaack, 1988, pp. 368-369); (Miller, 1985, p. 30)

[25] (Miller, 1985, p. 32)

[26] The Air Force subsequently directed ARDC to pursue such a bomber which ultimately led to the XB-70.

[27] Talbott has also previously stated on 10 March 1955 that "the B-58 program could be conducted if it did not cost more than $400,000,000."

[28] SAC supported the R&D decision but thought a 13 aircraft test program was larger than necessary.

[29] (Miller, 1985, p. 37)

[30] YRB-58A (55-0661, 55-0663, 55-0670, and 55-0671) were also assigned to the 6592nd TS at Carswell for R&D testing.

[31] YRB-58A (58-1011, 58-1012, 58-1015, and 58-1021 to 58-1023) were retained at CVCFW for corporate service tests. YRB-58A (58-1015) was also used for testing at the AFFTC.

[32] The term "deadman switch" referred to the Special Weapons Emergency Separation System (SWESS) which ensured that, once armed by the crew nearing the target, the pod (containing the nuclear weapon) would still drop if the crew was incapacitated or killed.

[33] (Miller, 1985, p. 44)

Models & Variants

[1] (Knaack, 1988, p. 365)

[2] (Miller, 1985, p. 46)

[3] (Knaack, 1988, pp. 378-379)

[4] (Brewer, 1994-2024)

[5] (Miller, 1985, p. 49)

[6] (Miller, 1985, p. 50)

[7] (Miller, 1985, p. 47)

[8] References including the Convairiety newspaper, dated 8 June 1960, say total hours were 179 even though the individual monthly hours only add to 175 hours.

[9] (Knaack, 1988, pp. 383-384)

[10] Delays in Category I and II testing led SAC to agree in March 1959 to move the start of Category III testing from October 1959 to March 1960. Further delays in the start date occurred as the aircraft control systems were modified.

[11] (Hirsch, 1964, p. 56)

[12] The previous longest known Mach 2.0 flight was logged by a Carswell B-58 flying from Seatle to Carswell in 70 minutes.

[13] Of the remaining six aircraft, five (58-1008, 58-1012, 58-1017, 58-1022, and 58-1023) were previously destroyed during testing and one (58-1007) was converted to TB-58A.
[14] (Knaack, 1988, pp. 352-353)
[15] (Brewer, 1994-2024)
[16] Selected project JUNIOR FLASH-UP aircraft also received the SENIOR FLASH-UP modifications.
[17] Built by Hughes Aircraft allowed B-58 aircraft to communicate 5,000 miles in any direction with the SAC worldwide communications network.
[18] The control system and structural modifications were implemented as a result of the YRB-58A (55-0664) crash on 7 November 1959.
[19] (Brewer, 1994-2024)
[20] Note the small size of the gauges despite that they were critical to flight because they managed all flight control systems.
[21] (Holt, 2023, p. 33)
[22] (Brewer, 1994-2024)
[23] (Holt, 2023, pp. 52-54)
[24] Low-level capability was planned for the B-58B but this aircraft was never built. Plans to implement into the B-58A were never fully carried out.
[25] (Hirsch, 1964, p. 55)
[26] (Holt, 2023, pp. 30-31)
[27] Some testing of the SACseat was also accomplished at the Supersonic Military Air Research Track (SMART) in the Mojave Desert at Hurricane Mesa, UT as was the Crew Escape Capsule.
[28] (Brewer, 1994-2024)
[29] (Hirsch, 1964, p. 53)
[30] (Brewer, 1994-2024)
[31] (Brewer, 1994-2024)
[32] (Holt, 2023, p. 58)
[33] (Hirsch, 1964, p. 53)
[34] The first human supersonic ejection of a capsule manned by Air Force volunteer Master Sergeant William C. Barbour was planned for May 1962 but was cancelled due to issues encountered with animal testing. An ejection at Mach 1.3 followed by one at Mach 1.6, both at 35,000 feet, were planned.
[35] The first manned low-level ejection was planned for August 1962 at Edwards with the YRB-58A (55-0661) flying at Mach 0.9 and 2,500 feet. However, this test was never completed and all testing was ultimately carried out with animals only.
[36] On 1 January 1963, the required time was changed from 15 minutes to 5 minutes.
[37] The first TB-58A (55-0670) was delivered on 13 August 1960.
[38] (Holt, 2023, p. 44)
[39] (Brewer, 1994-2024)
[40] (Brewer, 1994-2024)
[41] Redesignated Aerospace Defense Command in 1968.
[42] (Miller, 1985, pp. 109-110)
[43] (Brewer, 1994-2024)
[44] (Brewer, 1994-2024)

Armament and Weapons

[1] Most sources agree the maximum rate is 4,000 rounds per minute. However, according to "Convairiety" bi-weekly newspaper, dated 5 February 1958, the maximum firing rate was 7,000 rounds per minute. The National Museum of the U.S. Air Force website states a rate of 6,000 rounds per minute.
[2] (Hirsch, 1964, p. 57)
[3] (Miller, 1985, p. 104)
[4] The BA53 nuclear weapon was a smaller version of the Mk-53 with no fins that was able to fit in the weapons bay of the upper TC pod.
[5] On each flight only one of the four weapons was released on the target. Another 27 flights were planned to cover the B-58 flight spectrum but only two of the four weapons would be released on each flight.
[6] There is some speculation that the B-58 aircraft were painted in jungle camouflage for these tests, but this cannot be definitively verified.
[7] (Miller, 1985, p. 68)
[8] (Holt, 2023, pp. 143-144)
[9] (Miller, 1985, p. 107)

Organization and Basing

[1] In some cases, as many as 90 B-47 aircraft were assigned to one base.
[2] (Brewer, 1994-2024)
[3] The 43rd BMW leadership received notice on 14 May 1964 that it was to be moved to Little

Rock on 1 September 1964. The wing beat this date and was fully operational on 20 August.

World Records

[1] John Denver's father and first SAC pilot to fly the B-58.
[2] (Griswold, 1961, p. 182)
[3] The course was initially set at 666.7 miles (1,073 kilometers).
[4] Only 1 of 3 aircraft that attended Paris Air Show returned. B-58A (59-2443), named "Bye Bye Birdie", also crashed on 15-Jun-65 during landing for Paris Air Show at Le Bourget Airport, France. The aircraft was overweight due to excess fuel still onboard, crashed short of the runway, hit instrument approach beacons, and burst into flames. Pilot killed.
[5] (Miller, 1985, p. 61)

Operations

[1] (Hirsch, 1964, p. 52)
[2] (Hirsch, 1964, p. 58)
[3] (Hirsch, 1964, p. 58)
[4] (Hirsch, 1964, pp. 58-59)
[5] (Holt, 2023, p. 132)
[6] (Miller, 1985, p. 59)
[7] (Hirsch, 1964, p. 52)
[8] B-52 aircraft also underwent extensive structural modifications and repairs to the airframe to help mitigate the effects of turbulence on the airframe.
[9] (Holt, 2023, pp. 109-110)
[10] (Holt, 2023, pp. 54-55)
[11] Rocky had previously witnessed the crash and fire of B-58A (61-2063) on 26-Aug-63 in which two crew members, including the navigator/bombardier, died trying to escape the aircraft engulfed in flames by climbing over the wing. He said to another crew member "I'll never die that way", which could also explain why he chose to eject rather than chance the flames.

Displays and Disposition

[1] (Holt, 2023, p. 156)
[2] McNamara later explained that he delayed the FB-111A procurement announcement because he needed presidential approval. However, many in the Air Force and Congress believed that the decision to retire the older B-52 aircraft was dependent upon procuring the FB-111A. It still remains unclear why McNamara chose to delay the announcement.
[3]The SCAD program was soon cancelled and replaced by the Air Launched Cruise Missile (ALCM).
[4] (Hirsch, 1964, p. 47)
[5] B-52 aircraft also underwent extensive structural modifications and repairs to the airframe to help mitigate the effects of turbulence on the airframe.

About the Author

H.J. Campbell is a long-time veteran of the U.S. Air Force and the defense aerospace industry. He completed 20 years in the Air Force including 15 years in SAC. His SAC assignments included the 42nd BMW at Loring AFB, ME, 320th BMW at Mather AFB, CA, 43rd Strategic Wing at Andersen AFB, Guam, and HQ SAC Maintenance Standardization and Evaluation Team (MSET). After SAC was deactivated in Jun 1992, he was transferred to HQ Air Mobility Command (AMC) where he supported much needed KC-135 avionics upgrades. After his retirement from the Air Force, he served in the aerospace industry as a program manager and director for modification, maintenance, and logistics programs on KC-135, KC-10, C-130, C-27, and other Air Force aircraft. With this book, another in a planned series about SAC aircraft, he is pursuing his "third career" as a military historian and author.

Also available from *Electrikbooks* at Amazon.com, BarnesandNoble.com, and BooksaMillion.com

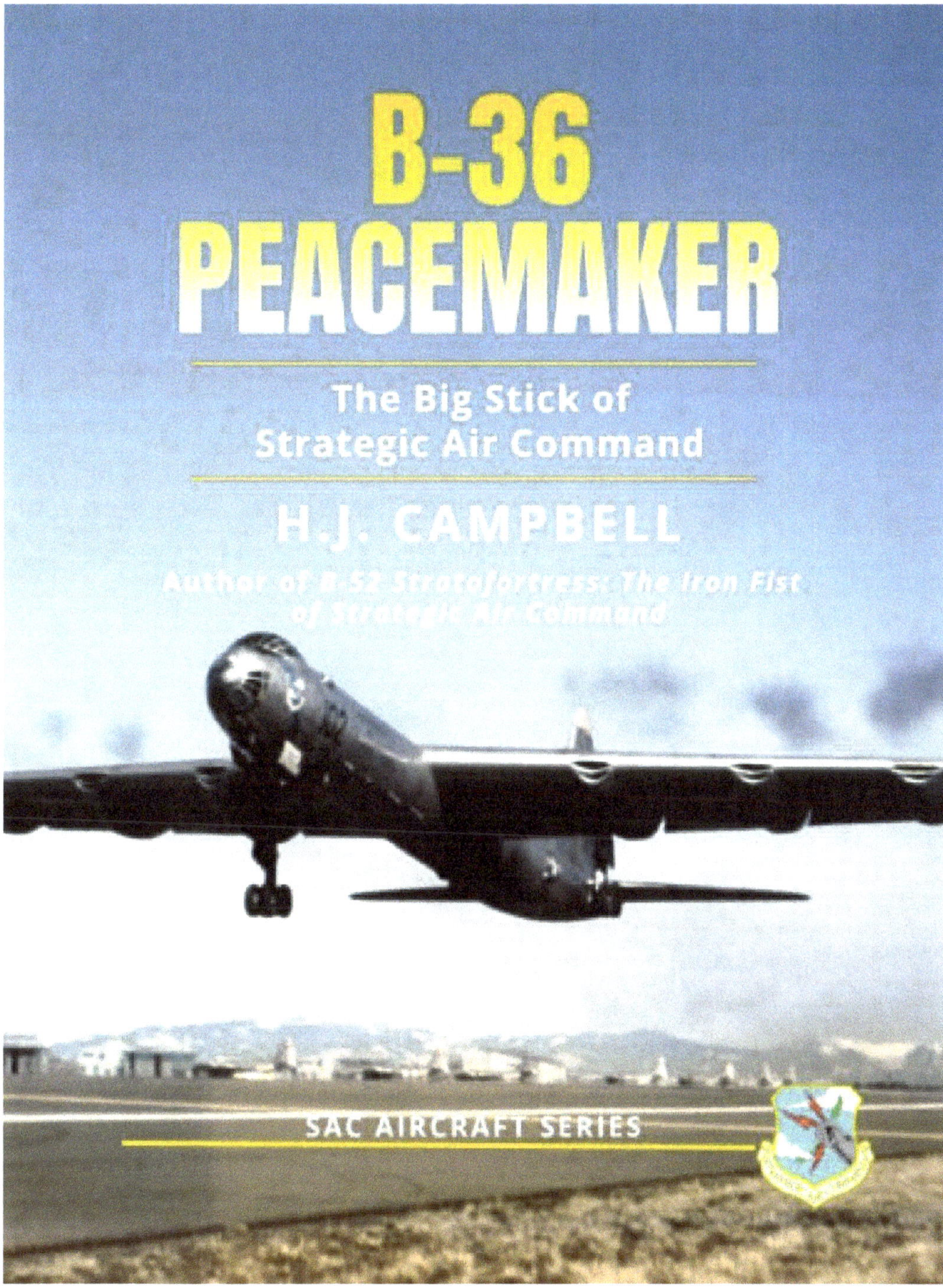
B-36
PEACEMAKER
The Big Stick of
Strategic Air Command
H.J. CAMPBELL
Author of B-52 Stratofortress: The Iron Fist
of Strategic Air Command
SAC AIRCRAFT SERIES

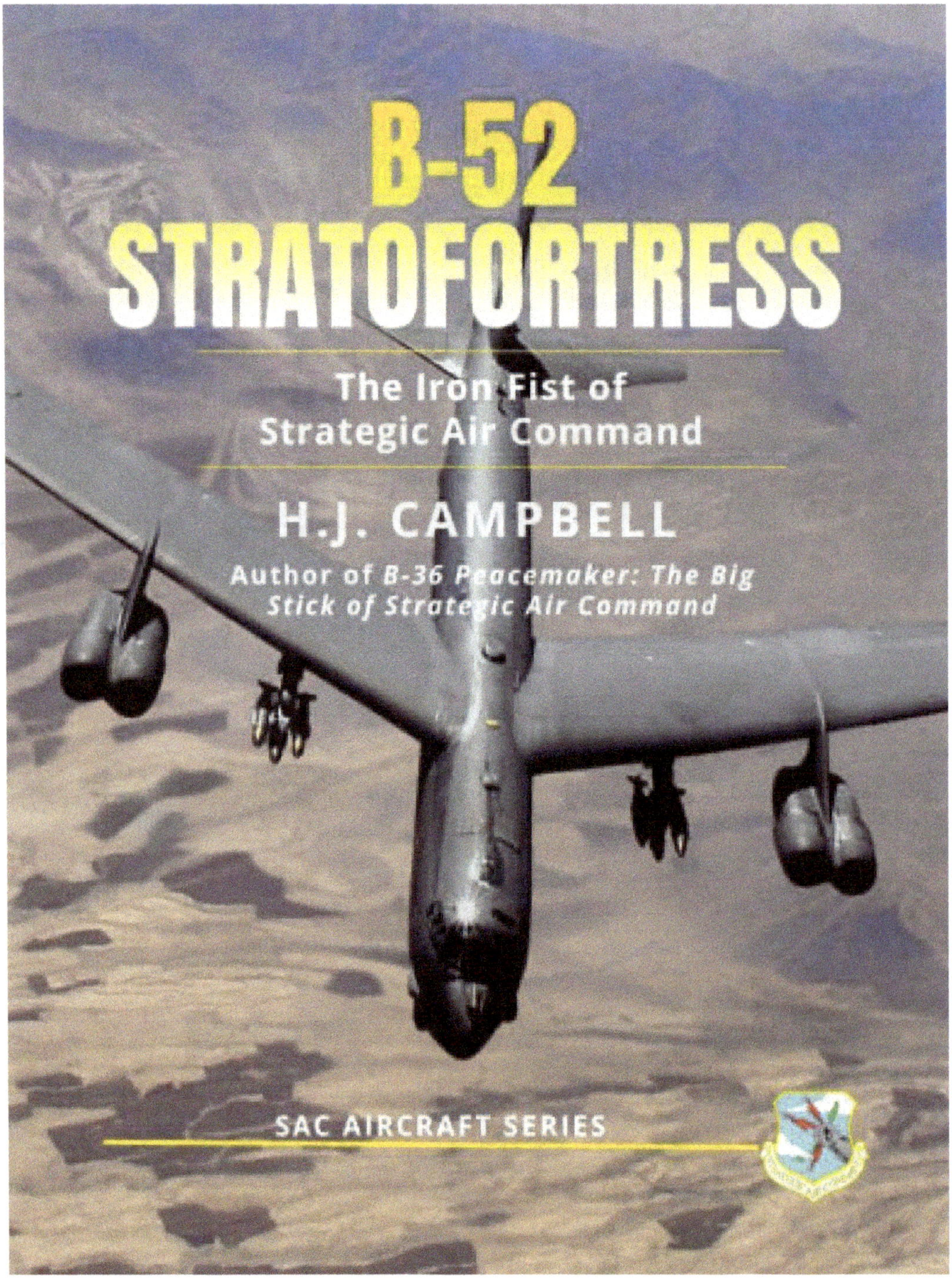
B-52
STRATOFORTRESS
The Iron Fist of
Strategic Air Command
H.J. CAMPBELL
Author of B-36 Peacemaker: The Big
Stick of Strategic Air Command
SAC AIRCRAFT SERIES

www.ingramcontent.com/pod-product-compliance
Lightning Source LLC
LaVergne TN
LVHW082103300926
849194LV00026B/294
* 9 7 8 1 7 3 7 4 9 8 2 8 5 *